Characterizing Groundwater Recharge: A Comprehensive Isotopic Approach

The mission of the Awwa Research Foundation (AwwaRF) is to advance the science of water to improve the quality of life. Funded primarily through annual subscription payments from over 1,000 utilities, consulting firms, and manufacturers in North America and abroad, AwwaRF sponsors research on all aspects of drinking water, including supply and resources, treatment, monitoring and analysis, distribution, management, and health effects.

From its headquarters in Denver, Colorado, the AwwaRF staff directs and supports the efforts of over 500 volunteers, who are the heart of the research program. These volunteers, serving on various boards and committees, use their expertise to select and monitor research studies to benefit the entire drinking water community.

Research findings are disseminated through a number of technology transfer activities, including research reports, conferences, videotape summaries, and periodicals.

Characterizing Groundwater Recharge: A Comprehensive Isotopic Approach

Prepared by:
Jean Moran, Ph.D.
Lawrence Livermore National Laboratory
Livermore, CA
and
Mikel S. Halliwell, P.E.
Alameda County Water District
Fremont, CA

Jointly sponsored by:
Awwa Research Foundation
6666 West Quincy Avenue
Denver, CO 80235
and
Alameda County Water District
P.O. Box 5110
Fremont, CA 94537

Published by the

Awwa Research Foundation

Advancing the Science of Water®

IWA Publishing

DISCLAIMER

This study was jointly funded by the Awwa Research Foundation (AwwaRF) and the Alameda County Water District (ACWD). This work was primarily performed by ACWD, with significant assistance from the U.S. Department of Energy through Lawrence Livermore National Laboratory, which is managed by the University of California. AwwaRF, ACWD, and Lawrence Livermore National Laboratory (U.S. Department of Energy) assume no responsibility for the content of the research study reported in this publication or for the opinions or statements of fact expressed in the report. The mention of trade names for commercial products does not represent or imply the approval of or endorsement by AwwaRF, ACWD, or Lawrence Livermore National Laboratory. This report is presented solely for informational purposes.

Published by IWA Publishing, Alliance House, 12 Caxton Street, London SW1H 0QS, UK

Telephone: +44 (0) 20 7654 5500; Fax: +44 (0) 20 7654 5555; Email: publications@iwap.co.uk; Web: **www.iwapublishing.com**

AwwaRF report number 90941F. Originally published by AwwaRF for its subscribers in 2003. IWA Publishing version published 2004.

ISBN 1 84339 849 4

TABLE OF CONTENTS

TABLES

FIGURES

FOREWORD

The Awwa Research Foundation is a non-profit corporation that is dedicated to the implementation of a research effort to help utilities respond to regulatory requirements and traditional high-priority concerns of the industry. The research agenda is developed through a process of consultation with subscribers and drinking water professionals. Under the umbrella of a Strategic Research Plan, the Research Advisory Council prioritizes the suggested projects based upon current and future needs, applicability, and past work; the recommendations are forwarded to the Board of Trustees for final selection. The Foundation also sponsors research projects through the unsolicited proposal process; the Collaborative Research, Research Applications, and Tailored Collaboration programs; and various joint research efforts with organizations such as the U.S. Environmental Protection Agency, the U.S. Bureau of Reclamation, and the Association of California Water Agencies.

This publication is a result of one of these sponsored studies, and it is hoped its findings will be applied in communities throughout the world. The following report serves not only as a means of communicating the results for the water industry's centralized research program but also as a tool to enlist the further support of the nonmember utilities and individuals.

Projects are managed closely from their inception to the final report by the Foundation's staff and large cadre of volunteers who willingly contribute their time and expertise. The Foundation serves a planning and management function and awards contracts to other institutions such as utilities, universities, and engineering firms. The funding for this research effort comes primarily from the Subscription Program, through which water utilities subscribe to the research program and make an annual payment proportionate to the volume of water they deliver and consultants and manufacturers subscribe based on their annual billings. The program offers a cost-effective and fair method for funding research in the public interest.

A broad spectrum of water supply issues is addressed by the Foundation's research agenda: resources, treatment and operations, distribution and storage, water quality and analysis, toxicology, economics, and management. The ultimate purpose of the coordinated effort is to assist water suppliers to provide the highest possible quality of water economically and reliably. The true benefits are realized when the results are implemented at the utility level. The Foundation's trustees are pleased to offer this publication as a contribution toward that end.

Water utilities operating production wells are at the focal point of increasing public and regulatory concern regarding vulnerability of wells from pathogens and other contaminants derived from surface waters serving as sources of groundwater recharge. Developing reliable estimates of dilution effects and travel times as measures of well vulnerability is often difficult due to spatial variations in aquifer properties and complicated hydraulics between recharge facilities and wells. This report showcases the advantages and limitations of isotopic methods, involving age dating and noble gas tracers, in overcoming uncertainties inherent in more traditional approaches to well vulnerability assessment.

Edmund G. Archuleta, P.E.
Chair, Board of Trustees
Awwa Research Foundation

James F. Manwaring, P.E.
Executive Director
Awwa Research Foundation

ACKNOWLEDGMENTS

The authors are indebted to the cooperation and participation of the following individuals that were involved in this project or provided technical assistance:

Lawrence Livermore National Laboratory

G. Bryant Hudson, Ph.D.	M. Lee Davisson
Carolyn J. Koester, Ph.D.	H. Wayne Culham
Gail F. Eaton	Joan Beiriger

Alameda County Water District (ACWD)

Marian Gonzalez, Patti McMahon, and

Albert Bayen	Steven D. Inn, P.E.	Michael Morton
Douglas Chun, P.E.	Michael Lanier	Bobby Ojha
Mary Lou Dennis	Alex Liang	Rick Parker
Alvaro Domínguez	John Marchand	James Reynolds, P.E.
Derek Hou	John A. McHugh, R.G., CHG	Joanne Wallis
Jim Ingle	Larry Moore	

The technical guidance and input provided by the Project Advisory Committee (PAC) and the AwwaRF manager for this project, Frank Blaha, P.E., were enormously helpful and are greatly appreciated. The PAC members were Bruce Macler, Ph.D.; David Rogers, Ph.D.; and Julie Sueker, Ph.D., P.H., P.E.

We would also like to take this opportunity to recognize the following ACWD managers without whose support this project would not have been possible: General Manager Paul Piraino, former General Manager James Beard, Engineering Department Manager Craig N. Hill, P.E., and Operations Department Manager Karl B. Stinson, P.E. Also, a very special thanks is extended to ACWD's Board of Directors: Arthur Lampert (President), John H. Weed (Vice-President), James G. Gunther, Judy C. Huang, and Martin L. Koller.

EXECUTIVE SUMMARY

INTRODUCTION

The purpose of the project was to demonstrate the value of age dating and isotopic tracers in characterizing the flow dynamics and water quality changes in a complex groundwater domain that includes artificial recharge facilities and municipal pumping wells. The techniques employed in this study may be helpful to water utilities that are interested in determining the age of water in their production wells or the time of travel between recharge sources and wells. Age and travel-time have been recognized as important criteria in determining the vulnerability of potable water wells to bacteria and pathogens, as well as accidental chemical leaks up gradient of production wells. Travel times can be used as a way to determine the required frequency of groundwater monitoring in a wellhead protection zone where potential hazards have been identified. In addition, water utilities may benefit from the efforts in this study to document changes in water quality as surface water percolates from recharge ponds and then migrates as groundwater to production wells.

With the goal of providing a reliable, high quality water supply, the Alameda County Water District (ACWD) artificially recharges approximately 30,000 acre-ft of water per year. ACWD serves a population of approximately 330,000 in the cities of Fremont, Newark, and Union City, California. Approximately 35 percent of ACWD's average annual total potable water production is supplied by groundwater. Knowledge of the hydrogeologic characteristics of the groundwater basin, and the parameters that govern groundwater flow, are therefore of critical importance to maintaining ACWD's operations. Significantly improved qualitative and quantitative understanding of groundwater flow rates, characteristics of groundwater flow, and effects of artificial recharge on water quality are the product of the study described in this report. The investigative tools applied during the study include environmental isotope data, water chemistry, geochemical data, hydrogeologic data and artificial isotopic tracers. Application of isotope techniques represented a new direction for ACWD, and has provided invaluable information that could not have been acquired using modeling or conventional hydrogeologic techniques.

The apex of the Niles Cone Groundwater Basin, an alluvial fan adjacent to the southeast corner of San Francisco Bay, served as the study area for this project. (Maps and a detailed description of the study area are provided in Chapter 2.) Within this area of investigation, ACWD operates artificial recharge facilities and two major wellfields. The Hayward Fault runs through the study area and acts as a barrier to lateral groundwater flow, effectively dividing the groundwater domain into two hydrologically independent sub-basins, which have been named the Above-the-Fault (AHF) and Below-the-Fault (BHF) sub-basins. Surface water is delivered to deep recharge ponds in each sub-basin by Alameda Creek that drains a large watershed upstream of ACWD. The aquifers are highly permeable but recharge from the ponds is restricted by sediment build-up in the ponds. Many of the ponds have maximum depths in excess of 70 feet with the deepest pond having a depth of 100 feet. Consequently, they exhibit lake-like behavior including stratification. The two wellfields in the study area were located in close proximity to the recharge ponds and to each other, but on opposite sides of the Hayward Fault.

SCOPE OF WORK

The first phase of work involved sampling selected groundwater wells for natural isotopes. Groundwater ages in various wells were determined by the tritium-helium method. Oxygen-18 and Deuterium were also measured in an attempt to differentiate between local versus imported sources of recharge. Carbon-13 and Carbon-14 content in the water were measured to provide some additional insight on the age of the water, and to infer geochemical processes that contribute to dissolved inorganic carbon in groundwater.

Tracer experiments followed the initial study. Noble gas tracers were added to surface water in the recharge facilities. Use of multiple tracers enabled tracking the migration of percolated groundwater from individual recharge ponds. Tracer concentrations in surface water were recorded during the dosing period and for three weeks thereafter. The progress of tracer migration in groundwater was tracked over a time period of one year, approximately.

Groundwater and surface water samples were collected for analysis of water chemistry and coliphage (bacterial virus). Samples for this characterization were collected at strategic points in time in an attempt to coincide with the migration of the tracer. The purpose of this timing was to gain insight on water quality changes in a mass of recharge water or groundwater as it flows from one point in the aquifer to the next. Of particular interest were changes in water quality as surface water from the pond percolates to the aquifer. Such changes were inferred by comparing water chemistry in the ponds with water sampled from the wells, particularly wells in immediate proximity to the ponds. Outside the scope of the project was direct characterization of water chemistry associated with the layer of fine-grained sediment (referred herein as the clogging layer) deposited on the bottom and banks of the ponds. The clogging layer may have an important role in the alteration of water chemistry during seepage from the pond to the aquifer.

RESULTS – INITIAL STUDY

Groundwater ages, determined by the tritium-helium age dating method, varied from less than one year in monitoring wells close to the recharge facilities to greater than 20 years in wells that are more remote from the ponds or screened in deeper aquifers. Mean groundwater ages in the AHF sub-basin average about 5 years, indicating relatively rapid transport of groundwater to the wellfield. In the BHF sub-basin, groundwater ages spanned from 3 to 31 years and tracked the layered stratigraphy, with water from deeper aquifers having older mean ages. These ages represent the mean residence time in the basin for water drawn into the wells. Environmental isotopes of Oxygen-18 and Deuterium indicate that groundwater in the study area has been derived mostly from the local creek watershed. These isotopes also provided some evidence of evaporation in certain recharge ponds with long residence times. Analyses of dissolved atmospheric gases indicate that excess air is not significantly entrained in recharge water that percolated through the vadose zone. Dissolved gas analyses also indicate that large components of 'pre-modern' (pre-1950) water are not drawn at the production wells. The lack of variation in calculated time-of-recharge temperatures among groundwater samples from various wells suggests that seasonal surface water temperature variations are not preserved in groundwater within the study area.

RESULTS – TRACER EXPERIMENTS

The tracer experiments were conducted after completion of the initial study. The lack of migration progress in the BHF sub-basin indicate that certain BHF ponds, previously thought to be an important source of recharge to the BHF wellfield, either do not supply the wellfield or do so in much longer time frames than previously thought. Water that is actually delivered on a short time-scale was determined to originate from a small pond adjacent to the Hayward Fault.

In the AHF sub-basin, tracers from all the ponds that were dosed reached the targeted AHF wellfield within the time frame of the study.

Interpretation of the observed trends in tracer concentrations in the AHF surface water and groundwater required close examination of hydrologic conditions. Hydrologic parameters tracked during the tracer studies included groundwater pumping, pond surface water elevations, and groundwater piezometric head. Surface water inflows to one pond were measured and seepage rates from the pond were calculated. These seepage rates were compared with groundwater heads, pond levels, and inflow rates to determine if any of these parameters have a potential influence on seepage rates. The hydrologic analysis indicated that groundwater movement is controlled primarily by wellfield pumping while percolation is controlled primarily by a clogging layer on the bottom and sides of the pond beneath the water surface. Percolation from the pond is believed to occur mainly at very shallow depths where the clogging layer is likely to have the smallest thickness and is subject to periodic drying and cracking.

The distribution of tracer concentrations in groundwater was consistent with the above-described hydrologic conditions. The distribution of tracer concentrations in the ponds did not follow the assumption of complete mixing- an assumption based on the absence of a detectable thermocline. Tracers were not added to the pond surface, but instead were injected at a significant depth. Subsequent monitoring after dosing indicated tracer concentrations deep in the ponds remained significantly higher than concentrations at shallow depths. It is hypothesized that tracers fluxing out of the pond and into the aquifer had to migrate upward and then horizontally outward to the shallow parts of the pond near the shoreline. Although no thermocline was present to indicate stratification, the vertical distribution of tracer suggests vertical mixing in the water column was not complete, and the flux of tracer out of the pond may have been controlled by vertical diffusion. Because of the relatively high elevation of the diversion pipes, surface water is received in the ponds at relatively shallow depths. In-flowing water may have percolated out of the ponds in short-circuit-like fashion instead of mixing first with deeper pond water.

Tracers were detected in wells from the AHF sub-basin rapidly after dosing. Breakthrough curves for the monitored wells had an initially steep upward curve, rising to a peak, then dropped off from the peak very slowly, giving a long tail. These long tails are attributed in part to the apparently long tracer residence time in the ponds. Slower movement in deeper portions of the aquifer could have also given long tails to the breakthrough curves.

A number of results indicate that there is a high degree of heterogeneity associated with the AHF sub-basin. They include low observed tracer abundances, apparent variations of flow rate with depth and/or horizontal location, and differences in tracer curves for different wells. All wells show mixing of tracer-laden water with water from different sources, or with water recharged during a time when tracer was not present. Cursory mass balance considerations suggest that that only a small fraction of the water recharged from AHF ponds reaches production wells within one year. A small fraction of the water travels quickly in preferential channels or layers, reaching one of the production wells in only 60 days. This time of travel corresponds to a flow rate of approximately 7,000 feet per year. The bulk of the tracer-tagged water moved more slowly, on time frames that may prove to be consistent with the water ages measured by the tritium-helium method.

RESULTS – WATER QUALITY CHANGES DURING TRANSPORT

Documentation of water quality changes during recharge and subsequent migration in the aquifer was attempted but not achieved with the desired level of precision. Tracer- tagged water did not reach most of the BHF wells in the time frame of the study; therefore, a comparison of water quality between pond water and most BHF wells within a single snapshot-in-time was not meaningful. In the AHF sub-basin, tracers reached nearly all the targeted wells in the study area but at very dilute concentrations relative to those observed in the recharge ponds after dosing. The monitoring wells' long screens, which apparently intercepted non-tagged water from deeper intervals within the aquifer along with shallow, tagged groundwater, probably gave rise to the high degree of dilution. Transformation or retardation parameters responsible for water quality changes in the process of migration could not be quantified because of this dilution and the limited number of samples collected for analysis of water chemistry.

Despite these shortcomings, some of the observed changes in water quality parameters are dramatic and consistent over the study area, suggesting that some generalizations can be made. Several improvements to water quality appear to take place during recharge over short times and distances. These include elimination of coliphage, a reduction in levels of Total Organic Carbon (TOC) and non-point source Methyl-tert-Butyl Ether (MTBE), and a large decrease in nitrate concentration. Coupled processes of dissolution of carbonate minerals and oxidative decomposition of organic matter at the sediment/water interface, where transition to a reducing environment takes place, are consistent with the changes in geochemical parameters that were observed. The only significant degradation of water quality is an increase in some major elements that comprise total dissolved solids.

CONCLUSIONS AND RECOMMENDATIONS

Over the course of the study, several lessons were learned which could be used to improve experimental design and outcome for future research projects that may take place in the same study area. The ponds were found to be poor candidates for delivering tracer that can be reliably quantified (in terms of C/C_0) down-gradient. Substantial dilution, uneven mixing, and complicated hydraulics associated with the flux of pond water to the aquifer, gave rise to breakthrough curves that were not amenable for direct calculation of aquifer parameters. Volatilization of tracer from the ponds could not be quantified, which limited conclusions based on a comparison of tracer mass in pond water with that in groundwater. A better approach for such a purpose would be to inject tracer at a well(s) near recharge areas. Nevertheless, dosing the recharge ponds proved to be useful toward understanding the importance of recharge pond characteristics on groundwater quality.

Monitoring points with the characteristics necessary to allow quantitative analysis of flow and changes in water quality were also lacking in the present study. Specifically, nested monitoring wells with short, shallow screens, could be used to examine variations in flow velocity with depth, and could be used to capture water samples that are representative of a single source and recharge time. It would have also been useful to monitor surface water for tracer over a longer period of time following dosing. This would have enabled a better accounting of the tracer mass over time and a better explanation of the breakthrough curves. Specifically, shorter-screened wells would have helped determine if the long, gradually sloping tails of the breakthrough curves were a consequence of slow movement of tracer in deeper portions of the aquifer or, alternatively, a slow rate of tracer mass flux from the ponds. Improvements could also be made to the approach used in this study to characterize surface water and groundwater chemistry. Multiple samples over time should have been collected from the ponds and the wells to enable development of concentration-versus-time plots that could be compared with tracer breakthrough curves. In addition, longer-term monitoring is recommended to account for seasonal fluctuations in concentrations of the various naturally occurring constituents of interest.

For the Alameda County Water District, this study has shown that AHF surface water supplies need to be protected from contamination to the greatest extent possible, given the relatively fast travel time for a portion of the water from the ponds to production wells. Most of the BHF wells are not influenced by recharge water from the large BHF ponds on a short (one year or less) time frame.

Results presented here can also be used to make informed decisions regarding pond operations and effects of recharge on water quality. For example, scraping sediments from the recharge ponds will undoubtedly lead to faster recharge, but may compromise the observed improvements in water quality that take place as water moves to a chemically reducing environment. Dissolution of aquifer minerals should be considered given the possibility of structural changes to the sediments near the sediment/water interface.

Field-scale studies investigating the fate of highly contaminated surface water through pond sediment would have to be conducted in a less-sensitive study area where groundwater is not intended for potable use. Surface water in this study area had low levels of nitrate, TOC, and coliphage. Consequently, it would be problematic to use the results from this study to predict the extent of attenuation in the sediments that would occur if the surface water quality were typical of urban runoff or wastewater treatment plant effluent.

CHAPTER 1
INTRODUCTION

Artificial recharge of surface water into groundwater aquifers is becoming a common means to store significant quantities of drinking water in the western United States. Winter surface water flows are captured and recharged through surface spreading facilities (typically abandoned gravel quarries) followed by summer-time withdrawal from groundwater wells located within short distances (< 1 mile) from entry points. Furthermore, it is becoming increasingly popular to supplement surface water with reclaimed wastewater as a source of artificial recharge. Among the regulatory concerns in such water management practices is the quality of the surface water recharged and the resulting quality of the water at the wellhead. Regulators, in particular, are addressing the fate of a host of water quality parameters including total organic carbon (TOC), pesticides, volatile organic compounds, nitrate, and trace metals such as zinc, copper, and lead during artificial recharge. Most importantly, though, regulators are concerned with the occurrence of pathogens in the surface water and their persistence in the groundwater environment. This issue is central to the new USEPA-proposed groundwater disinfection rules. Moreover, these regulatory concerns are not just limited to situations of intentional artificial recharge, but are applicable to other geographic areas in the United States where surface water (rivers, lakes, and reservoirs) directly influences groundwater production.

Published studies of fate and transport of pollutants from surface water to groundwater exist for situations of direct injection and river/groundwater recharge systems. For example, Roberts et al. (1982) published a comprehensive study on the fate of TOC and trihalomethanes during direct injection of reclaimed wastewater. They showed that many of these constituents have adsorptive retention in aquifer material and are prone to degradation over several months. They were also able to demonstrate biodegradation of some trace organics by monitoring TOC, chemical oxygen demand and dissolved oxygen levels in the recharged water.

Schwarzenbach et al. (1983) reported results from transport and fate studies of organic compounds infiltrated from rivers to the groundwater in Switzerland. They showed that many volatile organics (e.g., chloroform, 1,1,1-trichloroethane, trichloroethylene, and tetrachloroethylene) were persistent in the groundwater environment, whereas organics such as alkylbenzenes were biodegraded over short distances. Similar results were reported by Ahel (1991) for the Sava River/groundwater system in Croatia, but in addition, organics such as nonylphenol polyethoxylate surfactants and nitrilotriacetate showed significant biotransformation whereas ethylenediaminetetraacetate was persistent.

Subsequent studies at the Swiss site have shown that oxidation-reduction (redox) conditions in river water and, more importantly, in river sediment significantly affect the transport and fate of nutrients (nitrate) and metals (Mn, Zn, Cu, and Cd) in the local groundwater system (Von Gunten et al., 1991). This finding agrees well with a large body of literature concerning the cycling of metals and nutrients in surface waters and oceans. These studies have shown the importance of redox chemistry in the cycling of metals that exhibit multiple valence states in natural waters (such as Fe, Mn, As, Se, and Cr), and of metals that exhibit only one valence state in natural waters (such as Ni, Cu, Zn, Cd and Pb) (Hamilton-Taylor and Davison 1995, Salbu and Steinnes 1995). Johnson et al. (1999) found that precipitation and dissolution reactions are important and that leaching of F, Ba, and As was appreciable in soil-water interactions during artificial groundwater recharge. Even a short retention time in a biologically active surface recharge system may significantly influence the water quality of recharged groundwater. Biologic activity (and, to a lesser extent, photochemistry) controls redox chemistry, and is far more active in surface water and sediment pore water than in groundwater.

In addition, increasing focus on groundwater transmission of pathogens from human and/or animal contaminated surface waters recharging into groundwater supplies places an added concern on short and long-term water quality during artificial recharge. Yates et al. (1987) and Rehmann et al. (1999) have documented that viruses can have significant resistance to inactivation in groundwater environments and may migrate at distances up to several hundred meters in certain groundwater environments. Intensive bench and field scale experiments have been conducted to investigate physical and chemical factors that may affect pathogen fate and transport in porous media (e.g., Bales et al. 1993, Bales et al. 1995, Jin et al. 1997, McKay et al. 1993, Pieper et al. 1997, Scholl and Harvey 1992, Toran and Palumbo 1992, Woessner et al. 2001).

Although much has been written about application of isotopic methods for characterizing groundwater flow and recharge, there has been a lack of case studies documenting artificial introduction of isotopic tracers to investigate the fate of recharge derived from surface waters, especially deep recharge ponds converted from gravel quarry pits. The studies of river/groundwater interaction cited above inferred, generally, flow-paths and rates based on temperature data. Furthermore, the previous work mainly investigated transport in the groundwater over short distances (e.g., 0 to 120 meters from the river).

For this study, conservative, non-toxic isotopic tracers and age dating techniques were used to acquire detailed information on the direction and rate of solute transport in a small, dynamic sub-area of a larger regional groundwater system. This targeted sub-area encompasses two highly productive wellfields in close proximity to an active geologic fault and a complex of deep, lake-like recharge ponds. Tracers were added to selected recharge ponds and the subsequent migration of the tracers toward the wellfields was tracked. Concurrently, natural water quality parameters were measured in the ponds, production wells, and monitoring wells between the ponds and production wells. Water quality parameters for this study included general minerals, alkalinity, pH and other field parameters, trace metals, total organic carbon (TOC), low level organics (including MTBE), and coliphage (a virus of coliform bacteria).

The water quality portion of this study was conducted with a reconnaissance level of effort. Aside from the isotopic tracers, no constituents- including coliphage- were artificially introduced into groundwater or surface water for this study. It should also be noted that the surface water sources of recharge are relatively pristine and that wastewater is not used as a source of groundwater recharge in this subject study area. Water quality was monitored concurrently with tracer in the hope of discerning gross non-flow related effects (transformation and/or retardation) from flow-related (advection and dispersion) processes under unaltered conditions. For groundwater basin managers, tracer studies present a rare opportunity for such discernment. In addition to age-dating and tracer techniques, natural isotopic composition (Oxygen-18, Deuterium, Carbon-13, and Carbon-14) was characterized to enhance understanding of the origin of the groundwater and various processes affecting groundwater chemistry.

The recharge ponds in the study area were originally gravel quarry pits that were eventually flooded to serve as recharge ponds. By virtue of their size and depth, they exhibit lake-like behavior including seasonal thermal stratification. The ponds have undergone significant buildup of fine-grain sediment after years of operation as recharge ponds. In sharp contrast with this sediment layer, native soils that comprise the surrounding aquifer are predominately coarse-grained, gravelly, and highly permeable. Investigation of the potential for water quality changes as recharge water passes through the sediment layer and into the immediate aquifer zone was also among the goals of this study.

CHAPTER 2

DESCRIPTION OF GROUNDWATER BASIN AND STUDY AREA

Alameda County Water District (ACWD) provides potable water to a population of over 330,000 in the cities of Fremont, Newark, and Union City, California. ACWD's average annual total potable water production between 1997 and 1999 was 54,000 acre-feet, with groundwater comprising about 36 percent of this total. ACWD's jurisdiction covers approximately 100 square miles on the southeasterly side of the San Francisco Bay Plain between San Francisco Bay to the west and the range of hills, known as the Diablo Range, located several miles inland to the east (Figures 2.1 and 2.2).

The geographic area of the aquifer system yielding potable water to ACWD coincides roughly with ACWD's jurisdictional area. Geohydrology of the groundwater basin is well documented (DWR 1967, 1968, 1967; Kolterman 1993). The water-bearing sand and gravel units that form the aquifer system were deposited by streams flowing from the Diablo Range. The most important alluvial deposit is the Niles Cone, which was formed by the ancestor of present-day Alameda Creek. As indicated in Figure 2.1, the watershed feeding Alameda Creek is about six times the size of ACWD (and the groundwater basin). Between periods of alluvial deposition, sea levels in San Francisco Bay rose and fine-grained sediment settled out to form aquicludes, separating the sands and gravels into regional aquifer layers. The aquicludes are thick near San Francisco Bay, but thin out with distance inland, ultimately disappearing altogether east of the Hayward Fault (DWR 1968).

The Hayward Fault (Figure 2.2), which trends in a general northwest direction from north of Berkeley to southern Fremont, acts as a barrier (albeit slightly leaky) to lateral groundwater flow. Effectively, this fault hydraulically divides the Niles Cone groundwater basin into two separate sub-basins: the "Above Hayward Fault" (AHF) and "Below Hayward Fault" (BHF) sub-basins on the east and west sides of the Hayward Fault, respectively. The fault is not exposed at ground surface, but according to the California Department of Water Resources (DWR 1967), observations from deep excavations noted a "vertical clay dike composed of highly sheared and contorted clay." DWR suggests movement along the fault may have off-set the gravel aquifers and caused development of a "clay gouge," whereby scattered cobbles, pebbles, sands, and other unconsolidated materials were transformed at depth into relatively impermeable material.

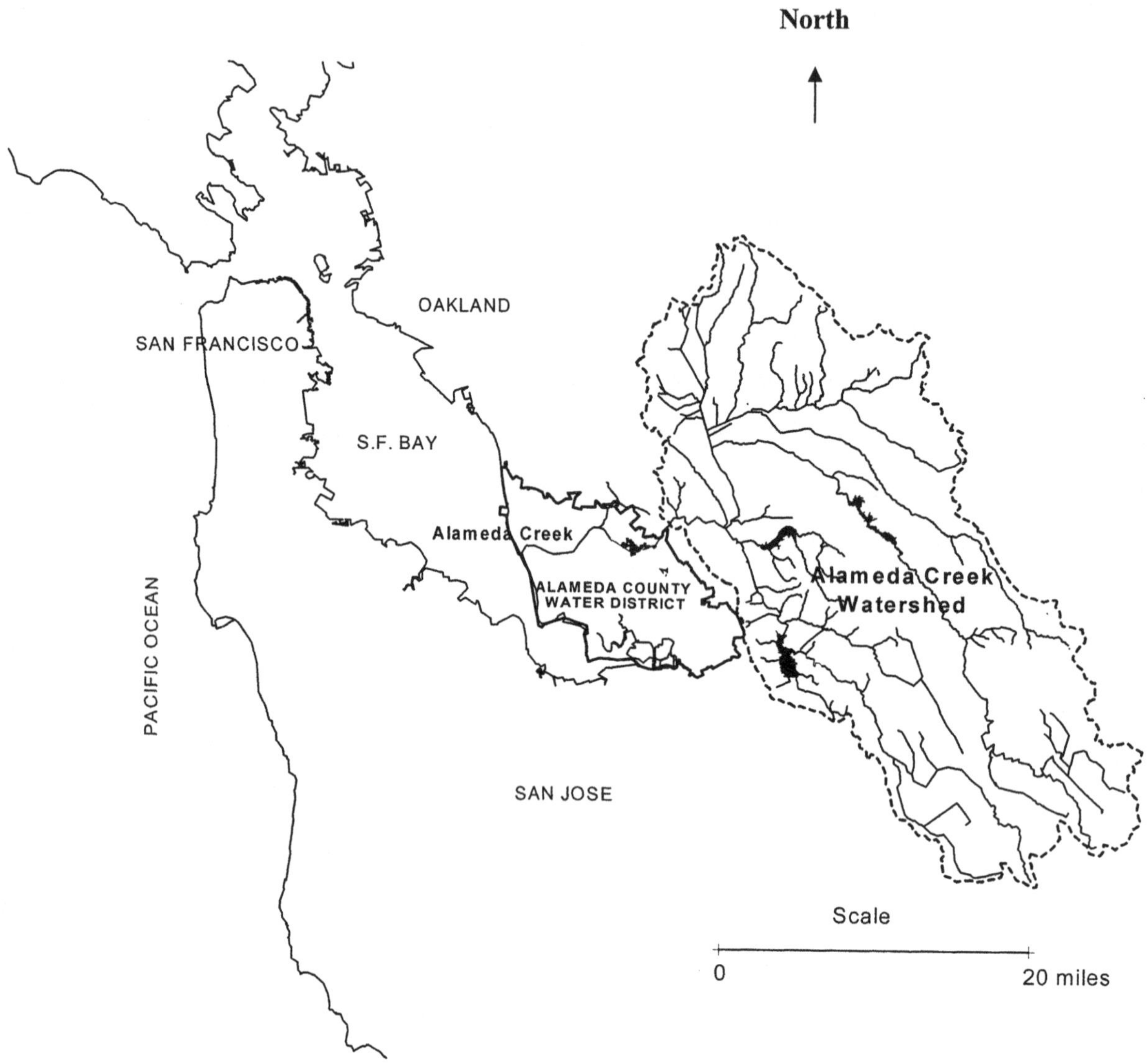

Figure 2.1 Location map: Alameda County Water District and Alameda Creek Watershed

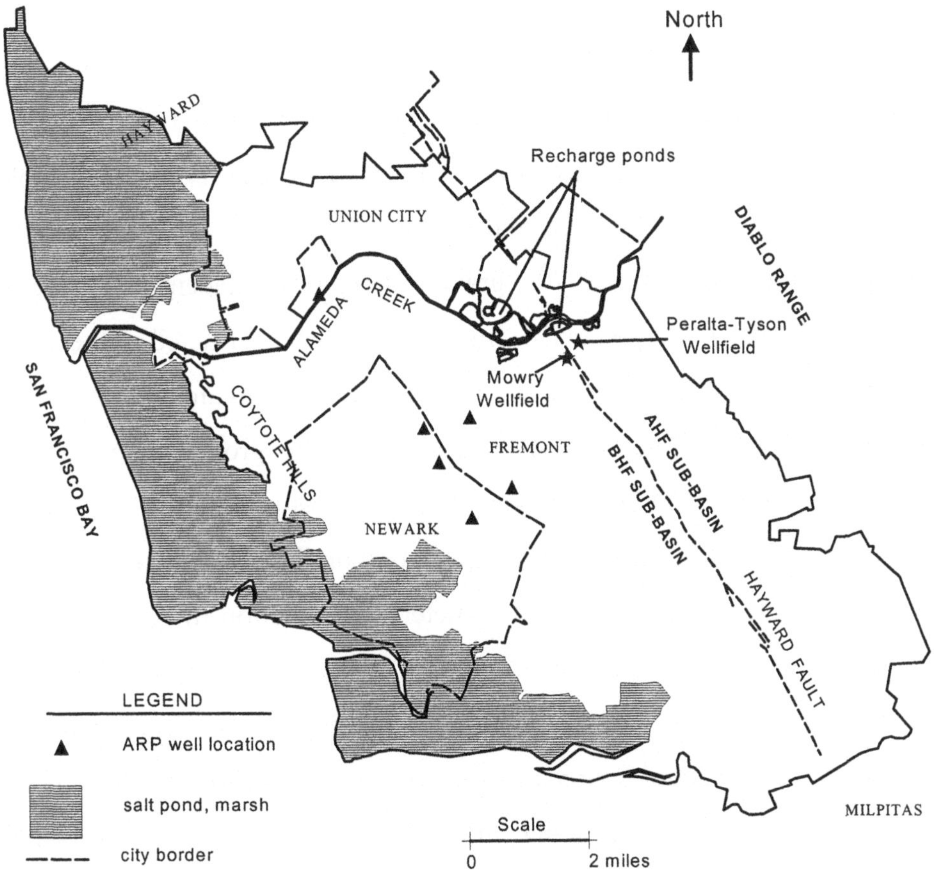

Figure 2.2 Alameda County Water District

Within the Niles Cone, the groundwater piezometric head immediately above the fault is typically 10 to 20 feet higher than the head immediately below the fault.

ACWD conducts artificial recharge and operates a major wellfield in each of the sub-basins. The recharge facilities are arranged in a contiguous complex that is comprised of Alameda Creek and adjacent recharge ponds. ACWD's two wellfields, the Peralta-Tyson Wellfield (AHF) and the Mowry Wellfield (BHF), are located close to the recharge facilities and to each other. Less than 1,800 feet separate the two wellfields; however, they do not exert a hydraulic influence on each other since they are on opposite sides of the fault. Similarly, AHF recharge facilities do not impact groundwater flow below the fault, and vice-versa.

The artificial recharge operation was established to ensure adequate replenishment of groundwater supplies in both sub-basins and to prevent and mitigate sea water intrusion in the BHF sub-basin. In the 1960s and 1970s, gravel quarry pits adjacent to Alameda Creek were converted by ACWD into recharge ponds. Inflatable dams, constructed in the 1970s and 1980s, impound water in Alameda Creek so that it can be diverted off-stream and into these ponds.

Quantities of the various sources and sinks to the groundwater basin are routinely calculated or estimated by ACWD on an annual basis. Most production wells in the District are metered. Monthly artificial recharge is estimated based on the metered amount of water diverted into recharge ponds, plus an estimate of non-metered percolation in Alameda Creek, minus an estimate of water evaporated from the ponds. ACWD uses *Integrated Groundwater Surface Model* (IGSM) (Montgomery Watson 1996), a time-dependent, finite-element groundwater model, to estimate other natural sources and sinks such as outflow to San Francisco Bay and gains from direct rain, applied water, and interflow from the hills. These components of the groundwater balance are further explained in subsequent sections and are quantified in Table 2.1 for the years coinciding with this study. Pumping and percolation from direct rain and applied water are broken down in the table to give separate values for AHF and BHF sub-basins. Artificial recharge, however, is difficult to differentiate according to sub-basin because Alameda Creek and certain individual ponds span the fault. The fraction of recharge to AHF is estimated by balancing the other inputs and losses with the known change in storage as provided by indicator wells. The remainder of the total is then assigned to the BHF sub-basin.

Table 2.1
Groundwater budget for ACWD 1997 - 1999
(all values in acre-feet)

Source/sink identification	Source of value	CY 1997	CY 1998	CY 1999
Above Hayward Fault (AHF) Sub-basin				
Peralta-Tyson (PT) Wellfield pumping	Measured	12,200	10,600	14,200
Private well pumping	Measured	1,100	800	700
Loss across the Hayward Fault to BHF	IGSM-estimated	0	200	-100
Artificial recharge	Fraction of total* is estimated	7,900	12,200	7,400
Percolation from direct rain, interflow from hills, and irrigation	IGSM-estimated	2,300	4,400	1,800
Below Hayward Fault (BHF) Sub-basin				
Mowry Wellfield pumping	Measured			
1. Newark Aquifer wells		3,000	3,000	2,200
2. Centerville-Fremont wells		4,700	3,400	4,300
3. Deep Aquifer wells		100	0	0
ARP well pumping (all aquifer layers)	Measured	5,700	9,900	7,600
Private well pumping (all aquifer layers)	Measured	3,400	2,300	2,100
Gain from flow across Hayward Fault	IGSM-estimated	0	200	-100
Outflow to San Francisco Bay, salt marshes and ponds	IGSM-estimated	2,700	8,600	8,400
Other outflow	Estimated	0	11,400	14,000
Artificial recharge	Fraction of total* is estimated	5,700	30,300	26,000
Percolation from direct rain, interflow from hills, and irrigation	IGSM-estimated	11,000	18,200	10,200

*total recharge is determined by adding the amount diverted into ponds (a measured value) and the amount percolated in Alameda Creek upstream of the rubber dams (an estimated value).

ABOVE HAYWARD FAULT HYDROGEOLOGY

Geographically, the Above Hayward Fault sub-basin occupies approximately 9 square miles, all of which are in Fremont. The land use in the sub-basin is almost entirely residential and commercial. With the Hayward Fault to the west and hills to the north, south, and east, the AHF sub-basin is effectively contained by no-flow boundaries. The most productive portion of the AHF sub-basin is estimated to be in the northern third, which includes ACWD's Peralta-Tyson (PT) Wellfield and the AHF-portion of ACWD's recharge operation (included in the area of interest for this study). This assessment is based on a report by the California Dept. of Water Resources (DWR 1967) which includes a description of the changes in permeability and well specific capacity. According to this reference, specific capacities in the northern third of the sub-basin range from 200 to 400 gpm/ft. Moving to the south, specific capacities drop quickly down below 100 gpm/ft, and are below 10 gpm/ft in the southern extreme of the sub-basin. This trend in decreasing specific capacity coincides with distance from Alameda Creek.

In addition to artificial recharge, sources of groundwater to the AHF sub-basin include shallow subsurface flow from the hills and deep percolation of direct rainfall and irrigated water. As only a small amount of groundwater likely leaks across the fault, potable production at the PT Wellfield is considered to be the most important sink in the AHF basin. Table 2.1 provides a breakdown of ACWD's estimated groundwater budget for the AHF sub-basin.

The level of surface water in AHF recharge facilities ranges from 44 to 67 feet mean sea level (MSL). Groundwater levels in the sub-basin are typically between 20 to 40 feet MSL.

The AHF sub-basin is generally considered as having only a single major aquifer unit from ground surface to bedrock. However, a thin layer of fine-grained soil under the PT Wellfield at a depth of 115 feet below ground surface (about 80 feet below the phreatic surface) may act as an aquitard.

BELOW HAYWARD FAULT

Hydrology

Geographically, the BHF sub-basin occupies approximately 90 square miles, including 10 square miles in southern Hayward (outside ACWD's service area for potable water distribution) and 40 square miles of uninhabited marshes and salt evaporator ponds along San Francisco Bay. The remaining 40 square miles consist of residential, commercial, and industrial areas of Fremont, Newark, and Union City.

The sub-basin is comprised of three major regional aquifer layers that are separated from each other by the aforementioned clay aquicludes. The Newark Aquifer, the uppermost regional aquifer layer, starts from 0 to 40 feet below ground surface and extends to a depth of 140 feet. Beneath the Newark Aquifer, the combined Centerville and Fremont Aquifers extend from 180 to 390 feet, and the Deep Aquifer from 400 feet to an undetermined depth (DWR 1968). Because of the extent to which the Centerville and Fremont Aquifers are interconnected throughout the basin, these two aquifers are referred to collectively as the "Centerville-Fremont

Aquifer". Throughout the basin, the Centerville-Fremont and Deep Aquifers are confined. The Newark Aquifer is unconfined near the recharge area and the Mowry Wellfield (and the entire study area for this project) but is confined in other areas. In the confined areas of the Newark Aquifer, a clay cap, known as the Newark Aquiclude, extends from ground surface to the start depth of the sands and gravels that comprise the Newark Aquifer. In many areas within the sub-basin, groundwater is perched above the Newark Aquifer within non-regional sand and silt layers within the Newark Aquiclude.

All three major aquifer layers in the BHF sub-basin are hydraulically bounded to the east by the Hayward Fault. The Newark Aquifer thins out in the northern part of Union City and the southern part of Fremont, with the clays and silts effectively forming no-flow boundaries. Under the large complex of salt evaporator ponds and marshes along San Francisco Bay, the Newark Aquifer appears to be under the direct hydraulic influence of seawater with a constant head near 0 feet Mean Sea Level (MSL). The chloride concentrations in the Newark Aquifer vary from less than 250 ppm in inland areas to over 20,000 ppm under the salt evaporator ponds and marshes. Available chloride data for the southern portion of San Francisco Bay ranges between 10,600 and 18,700 ppm, generally decreasing in concentration with distance south (DWR 1968). Although these data date back to 1961, they remain relevant because they describe the chloride levels in bay water at the time of historical saltwater intrusion (discussed later in this section) into inland areas of the groundwater basin. Chloride levels in the Newark Aquifer near the salt ponds exceed bay water chloride concentrations; thus, it appears that the salt ponds have impacted groundwater. The salt evaporator ponds, which are actively operated by a salt producing company, have chloride concentrations ranging from bay water levels to incipient crystallization (250,000 ppm).

Like the Newark Aquifer, the Centerville-Fremont Aquifer is bounded to the north and south by low-permeable materials. To the west near San Francisco Bay, the Centerville-Fremont Aquifer also thins out considerably and becomes more fine-grained, but it may be transmissive enough to allow a small of amount of fresh water to flow under San Francisco Bay. The Deep Aquifer is effectively bound laterally by fine-grained deposits to the west and south, but extends northward and hydraulically interconnects with another regional aquifer unit underlying Hayward and other cities situated along the eastern shore of San Francisco Bay (DWR 1960, Luhdorff and Scalmanini Consulting Engineers 2002).

Since the end of the saltwater intrusion period (which ended in the late 1960s), the Newark Aquifer piezometric head generally has ranged between 5 to 20 feet MSL near the Mowry Wellfield, but approaches 0 feet MSL near the salt ponds and marshes adjacent to San Francisco Bay. Close to the Mowry Wellfield, the piezometric head in the Centerville-Fremont Aquifer is typically 5 to 15 feet less than that of the Newark Aquifer and the head in the Deep Aquifer is generally less than that of Centerville-Fremont Aquifer by 0 to 5 feet. Contour lines of equal groundwater head for the AHF Aquifer and all three major aquifer layers of the BHF sub-basin, interpreted from well levels measured in September 1998, are provided in Figures 2.3 (a), (b) and (c). The elevation of surface water in BHF recharge ponds fluctuates between 5 and 35 feet MSL.

Major sources and sinks to groundwater in the BHF sub-basin are quantified in Table 2.1 for the project study period. Sources include inputs from artificial recharge, interflow from the hills and percolation of rainfall and water irrigated onto land. A small amount of water may also be gained from the AHF sub-basin by horizontal leakage across the Hayward Fault. With exception of this fault leakage, which may be received by all three aquifer layers in the BHF sub-basin, the aforementioned sources are probably first received by the Newark Aquifer. Subsequently, the deeper Centerville-Fremont Aquifer receives water from the Newark Aquifer by leakage through the intervening aquitard. Similarly, leakage from the Centerville-Fremont Aquifer supplies the Deep Aquifer. In addition to pumping, sinks in the BHF sub-basin include Newark Aquifer outflow to San Francisco Bay, salt ponds, and marshes.

When the piezometric head of the Newark Aquifer, as measured in an indicator well approximately ½ mile west of the Mowry Wellfield, approaches 20 feet MSL, the amount of outflow appears to increase dramatically. This increase in outflow is apparent from the difficulty experienced by the District in achieving further increases in the amount of water storage in the BHF sub-basin (as measured by well levels) at this magnitude of head, despite continued recharge in excess of pumping. ACWD has not determined conclusively how the water is being lost. However, substantial upwelling of groundwater into Alameda Creek, just downstream of the artificial recharge facilities, has been observed when the indicator well reaches this critical level. Further recharge appears to be effectively resisted, as maintenance of high pond levels by continued inflow of surface water drives the upwelling of groundwater into segments of Alameda Creek immediately adjacent to or just downstream of the recharge ponds – like a short circuit. This groundwater sink is denoted in Table 2.1 as "other outflow." The IGSM

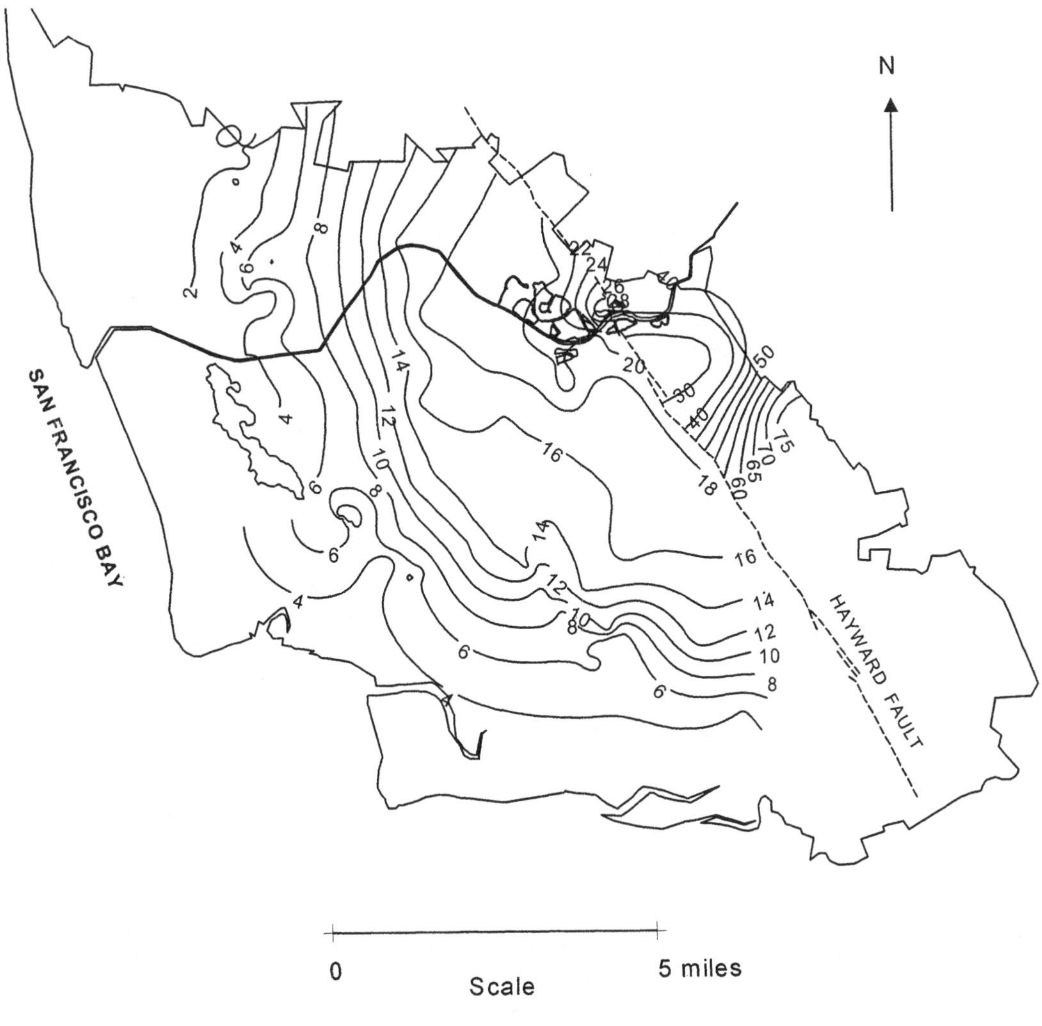

Figure 2.3a Piezometric head in Above Hayward Fault Aquifer (AHF) and Newark Aquifer (BHF) during Fall, 1998. AHF and BHF Sub-regions are east and west of the Hayward Fault, respectively.

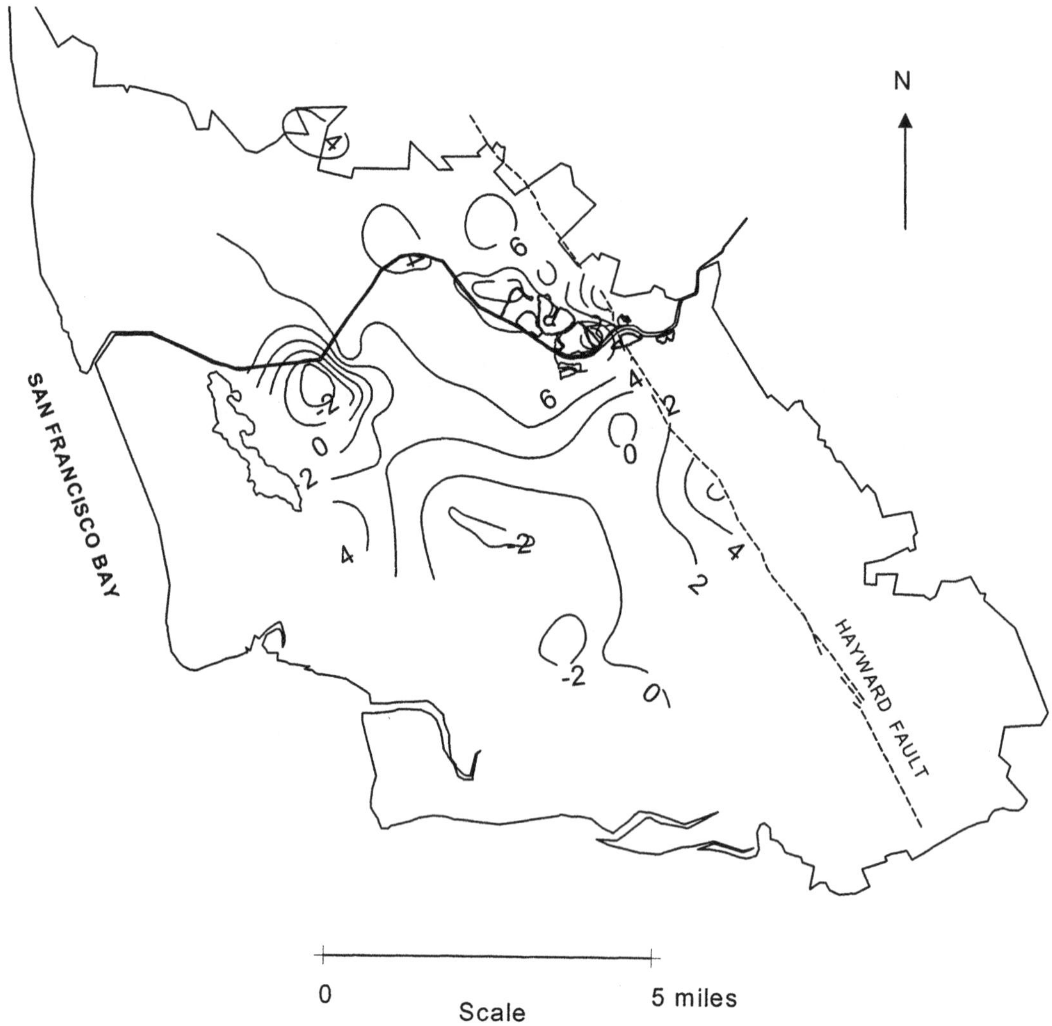

Figure 2.3b Piezometric head in Centerville-Fremont Aquifer (BHF) during Fall 1998

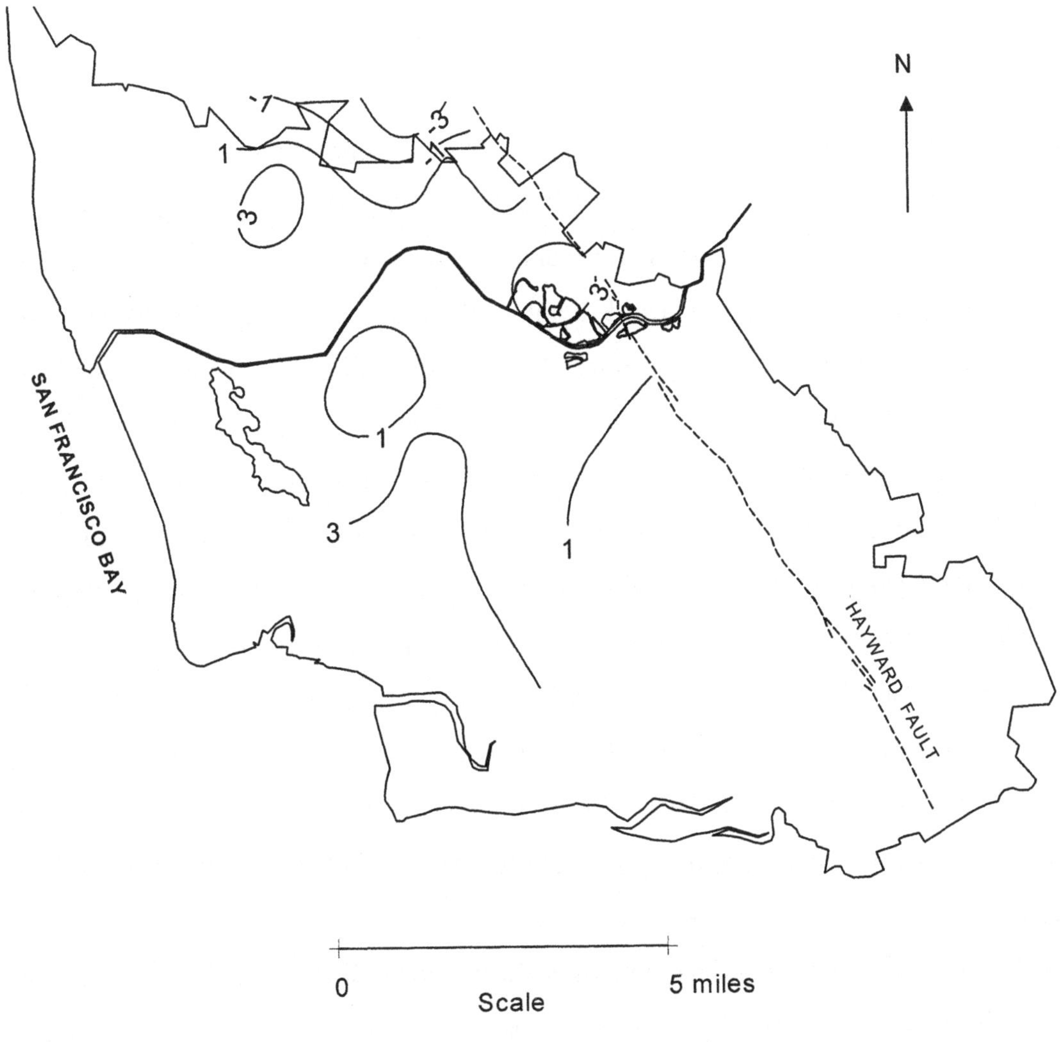

Figure 2.3c Piezometric head in Deep Aquifer (BHF) during Fall 1998

groundwater model was used to assist in the estimation of the quantity of "other outflow", as the amount of upwelling into Alameda Creek (and other possible loss mechanisms) has not been determined by more direct methods.

Sea Water Intrusion

Over-pumping and inadequate recharge from the early part of the 20th century through the 1960's caused the piezometric head of the Newark Aquifer to drop significantly below sea level, reaching a low of -60 feet MSL in 1963 (DWR 1960, ACWD 1997). Consequently, salt water from the San Francisco Bay, salt ponds and marshes entered the Newark Aquifer and migrated inland, ultimately reaching the Hayward Fault. Degraded water in the Newark Aquifer eventually migrated vertically to the Centerville-Fremont and Deep Aquifers. By 1974, the piezometric head in the Newark Aquifer was restored to above sea level, ending the long period of salt water loading into the groundwater basin. The flow direction has since been maintained toward the bay, and groundwater quality over a large inland area, including the study area for this project, has been restored to drinking water standards. However, significant plumes of brackish water remain in all three major aquifer layers of the BHF sub-basin.

To augment artificial recharge as a means of mitigating brackish water in the three major aquifer layers of the BHF sub-basin, ACWD installed nine brackish water extraction wells under its Aquifer Reclamation Program (ARP). The ARP wells are located in six sites (three sites have two wells which pump from different aquifers) in the western half of the basin (see Figure 2.2). Most are screened in the Centerville-Fremont Aquifer because brackish water in the Centerville-Fremont Aquifer (and Deep Aquifer), compared to the Newark Aquifer, is not as amenable to flushing with aggressive artificial recharge. Resistance to flushing stems from aquitards that separate the Newark Aquifer from the deeper aquifers. As mentioned previously, the aquitards are thickest at the western part of the basin near San Francisco Bay. Although the Centerville-Fremont ARP wells help remove brackish water, ACWD also operates them to counteract the tendency for production wells at the Mowry Wellfield to induce additional landward movement of brackish water in the Centerville-Fremont Aquifer. ARP pumping constitutes a significant demand on the groundwater basin, as indicated in Table 2.1.

IMPACT OF BASIN HYDROLOGY AND OPERATION ON THIS PROJECT

The amount of water flowing through the study area (i.e., to the Mowry Wellfield and Peralta-Tyson Wellfield) is dictated by basin hydrology, recharge, and pumping operations described above. Much of the recharge to the BHF sub-basin appears to flow toward the distal portions of the Niles Cone instead of the Mowry Wellfield. Recharging the BHF sub-basin in excess of potable water extractions may seem wasteful; however, it is helping to prevent seawater intrusion and to improve water quality in the brackish portions of the BHF sub-basin for future beneficial use.

In contrast to the BHF sub-basin, most of the water recharged to the AHF sub-basin appears to flow to production wells (mainly the Peralta-Tyson Wellfield).

STUDY AREA

Consistent with the project's objectives, the area of interest is that which extends between the recharge ponds and the two major wellfields. As the PT (AHF) and Mowry (BHF) Wellfields are only within 1,200 feet and 2,500 feet, respectively, from the nearest recharge pond, the study area is quite small relative to the size of ACWD's overall groundwater basin. However, the area includes some of the basin's most important sources and sinks. Consequently, the flow system being investigated is dynamic and complex. A map of the study area, including recharge ponds, production wells, and monitoring wells, is provided in Figure 2.4. Figure 2.5 provides a more detailed view of facilities involved in the tracer experiments, and includes the alignment of cross-sectional lines A-A' and B-B' for reference to Figures 2.6 and 2.7.

Local Hydrogeology

Geologic cross-sections A-A' (Figure 2.6) and B-B' (Figure 2.7) depict the significant stratigraphic units and the location of the phreatic surface of unconfined groundwater for selected alignments within the AHF and BHF portions of the study area, respectively.

In the AHF, a thin layer of fine-grained soil appears to extend from the PT Wellfield toward Kaiser Pond. It is not known the extent to which this potential aquitard is continuous over the rest of the AHF portion of the study area. Groundwater is generally unconfined

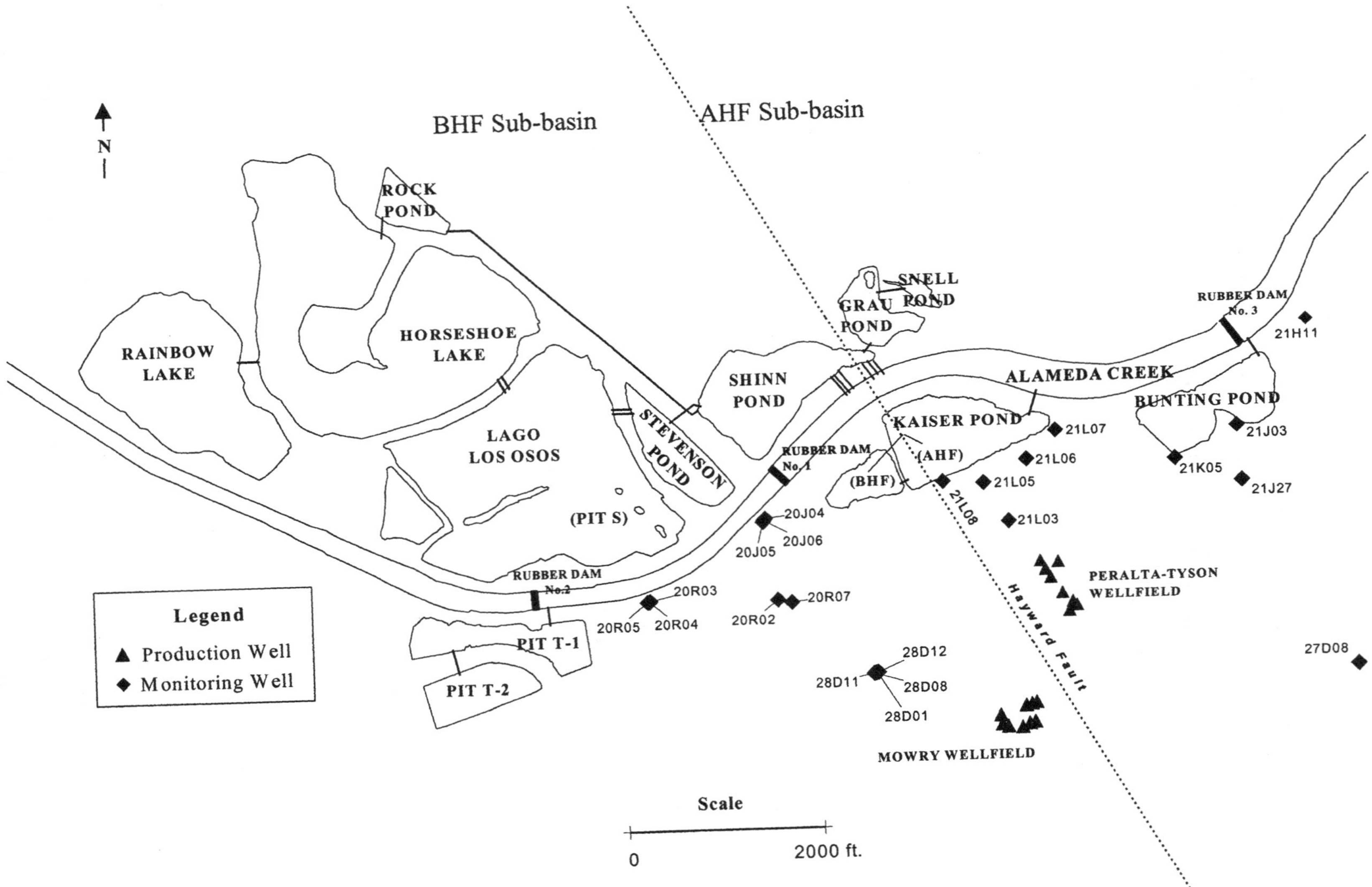

Figure 2.4 Study area

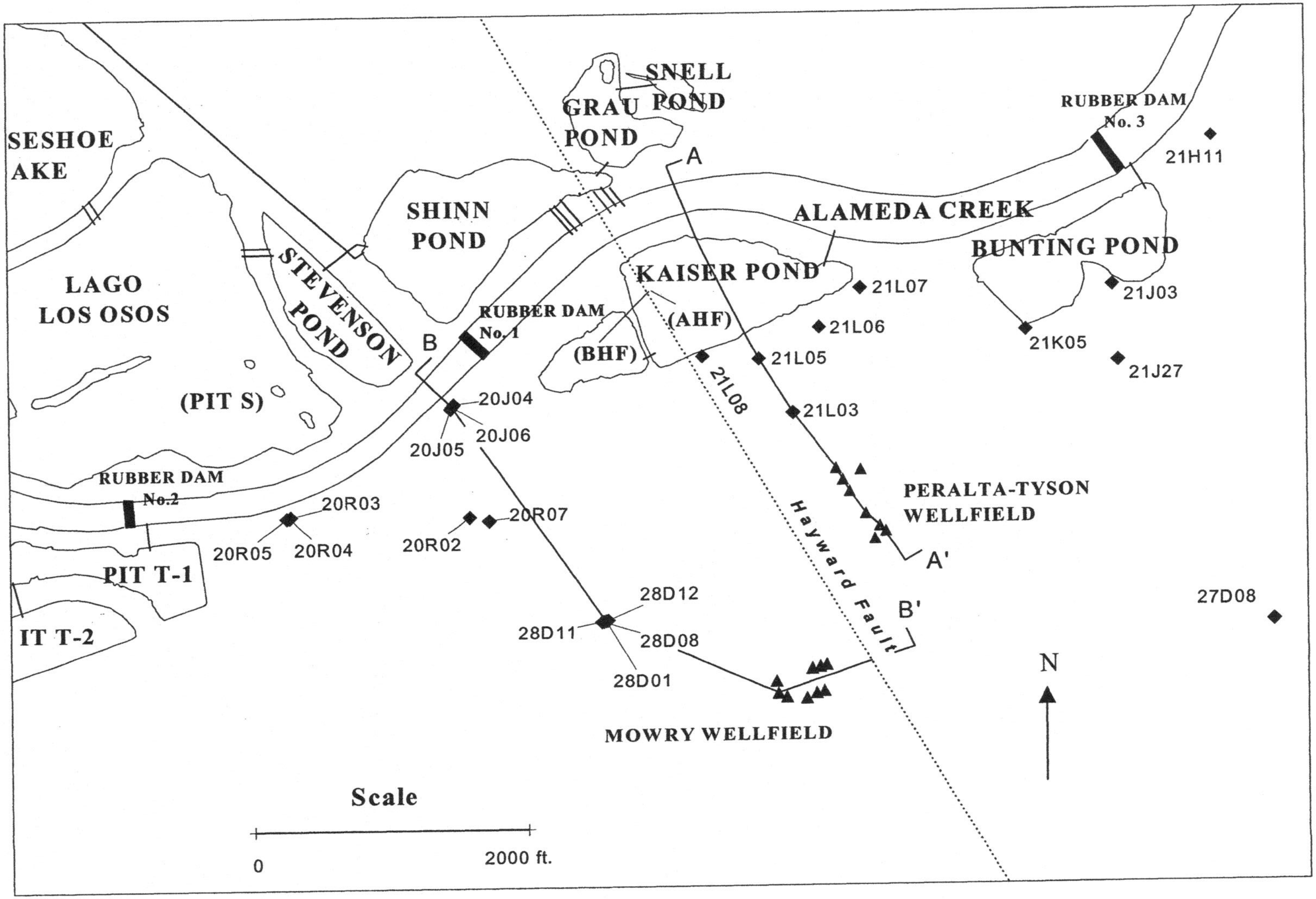

Figure 2.5 Horizontal alignment of cross-sections A-A' (AHF) and B-B' (BHF)

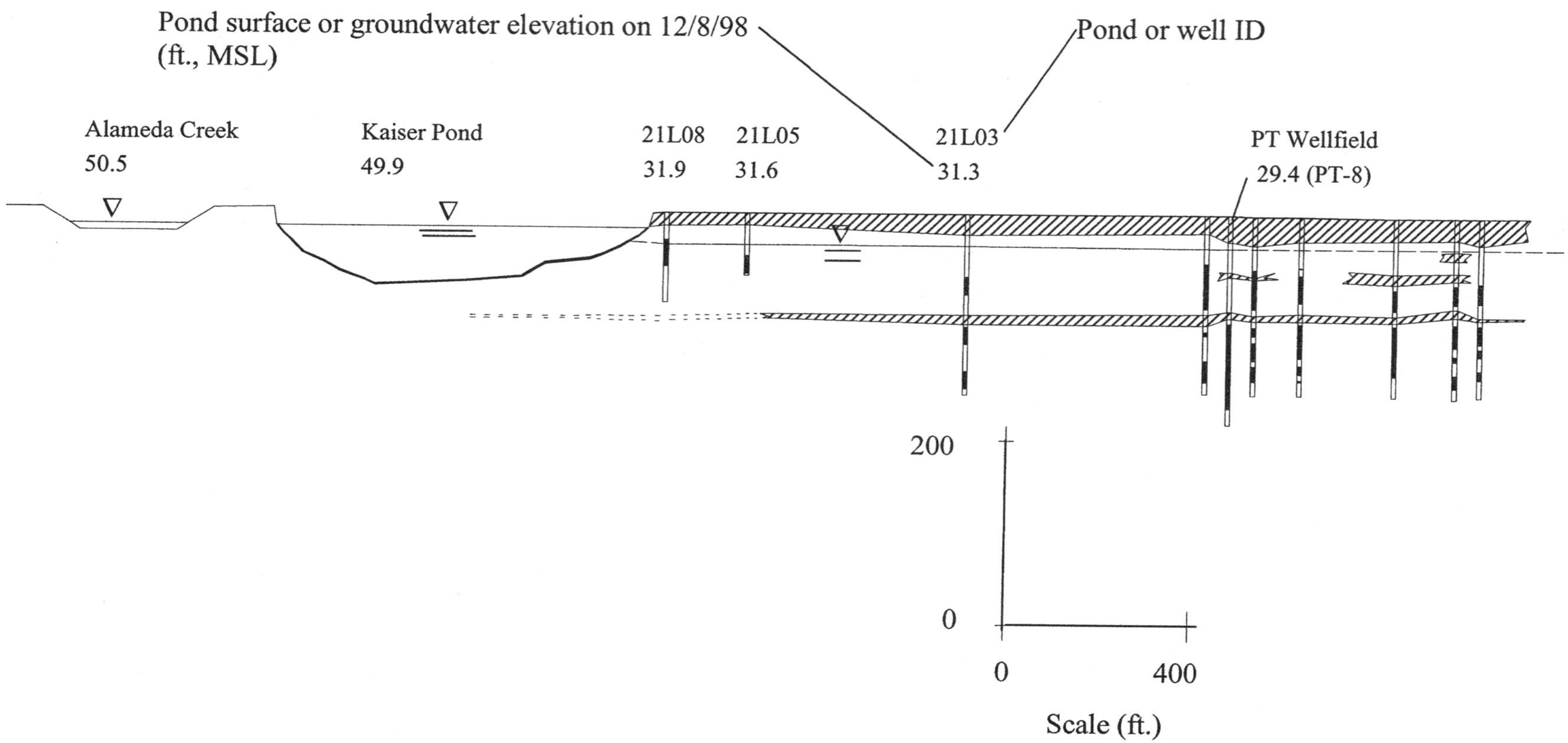

Figure 2.6 Profile of stratigraphy and piezometric head along A-A' (in AHF), from Alameda Creek to PT Wellfield. Shaded (cross-hatching) indicates fine-grained soil (clay and/or silt) and non-shaded areas represent coarse-grained soil. Stratigraphy at depths greater than 200 ft. below ground surface (b.g.s.) not shown. Depth to bedrock approx. 500 ft. b.g.s, according to DWR (1967).

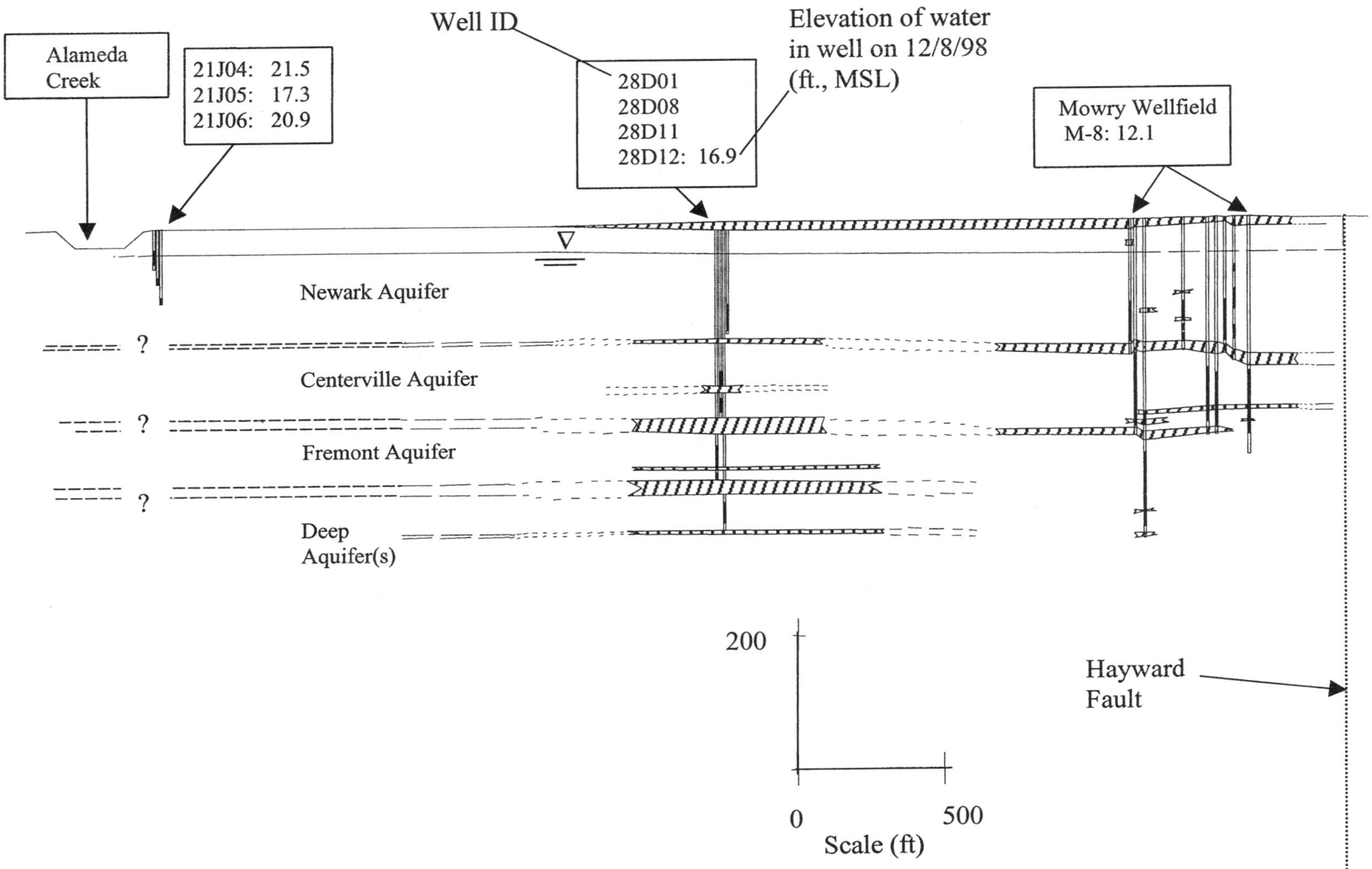

Figure 2.7 Profile of stratigraphy and peizometric head along B-B' (BHF), from Alameda Creek to Hayward Fault. Shaded (cross-hatching) indicates fine-grained soil. Dashed lines indicate the estimated location of aquitards away from deep wells where stratigraphy was confirmed. Depth to bedrock approx. 500 feet b.g.s., according to DWR (1967).

between the PT wells and the recharge ponds (and probably elsewhere within the AHF sub-basin); however, with reference to the finer-grained top soil zone, confined conditions may be possible with peizometric heads of 45 feet MSL or greater. However, such peizometric heads have rarely, if ever, been reached.

As indicated in the cross-section B-B', regional aquifer units of the BHF sub-basin are discernable from boring logs of Mowry production wells and the 28D-series monitoring well cluster. The lack of a confinement in the Newark Aquifer is typical of the forebay, which encompasses the study area.

Artificial Recharge Facilities

It is likely that the recharge ponds contribute a great deal to the hydraulic complexity. Once abandoned gravel pits, the recharge ponds are irregular both in shape and in orientation relative to each other. Moreover, many of them (but not all) were excavated to depths substantially below current operating groundwater levels. Data on the bathymetry and effective storage of the recharge facilities are provided in Table 2.2. In contrast to the ponds, Alameda Creek has a geometrically regular cross-section and slope, having undergone channelization work by the Army Corps of Engineers in the 1960s. Also, unlike the recharge ponds, the elevation of the bottom of the creek (segment within the study area) is typically (but not always) higher than the piezometric head of underlying groundwater. The creek is not lined with impermeable materials. The bottom is sand and gravel or grass-covered, and the sidewalls are riprapped with large boulders.

Kaiser Pond-AHF, Bunting Pond and adjacent and upstream segments of Alameda Creek are responsible for most the recharge to the AHF sub-basin. Smaller AHF ponds include Grau Pond and Snell Ponds on the north side of the creek. Shinn Pond spans the Hayward Fault but most of its area falls on the BHF side. Alameda Creek water is diverted into Kaiser and Bunting Ponds when Rubber Dams 1 and 3, respectively, are raised (Figure 2.4). Shinn pond is equipped with diversion pipes that enable it to collect water impounded by both Rubber Dams 1 and 3. Grau and Snell Ponds receive water from Shinn Pond.

Below the fault, there is a very large system of interconnected ponds known as the Quarry Pits. The Quarry Pits consist of Stevenson Pond, Lago Los Osos, Horeshoe Lake, Rock Pond, and Rainbow Lake (Figure 2.4). The remaining BHF ponds, located outside the Quarry

Table 2.2

Recharge facility data

Facility	Area of water surface at max. level	Vol. of water at max. level	Water level eleva.		Avg. eleva. of bottom	Eleva. of bottom at lowest point	Max. opera. level	Seepage rate*
			Low	High				
	(acres)	(ac-ft)	(ft. MSL)	(ft. MSL)	(ft. MSL)	(ft. MSL)	(ft. MSL)	(ac-ft/d)
Above Hayward Fault								
Rubber Dam 3	23	153	53	66	59	53	66	12
Bunting Pond	14	316	35	63	44	25	63	12
Grau Pond	6	96	35	46	30	29	46	2
Snell Pond	3	27	35	44	35	29	44	2
Kaiser Pit AHF	24	1,080	35	50	5	-20	50	8
Total AHF	70	1,671						36
Below Hayward Fault								
Kaiser Pit BHF	7	210	35	50	20	2	50	4
Shinn Pit	25	1,000	35	53	14	-10	54	6
Rubber Dam 1	45	300	43	53	45	43	53	8
Rubber Dam 2	12	35	30	35	31	30	35	10
Stevenson Pond	16	672	25	41	0	-20	35	6
Lago Los Osos								
Pit R2			0	35	-20	-43	35	2
Pit R3			0	35	-5	-15	35	2
Pit S1			0	35	-20	-46	35	2
Pit S2			0	35	10	6	35	2
Pit R2			0	35	-20	-43	35	2
total	90	3,400						10
Rock Pond	5	140	0	35	0	-13	35	4
Horseshoe Lake								
Pit G	48	2,300	0	35	-30	-62	35	10
Pit H	65	2,200	0	35	-20	-46	35	6
total	113	4,500						16
Rainbow Lake	46	2,700	0	35	-35	-69	35	10
Pit T1	12	348	10	24	-5	-10	24	2
Pit T2	12	348	10	24	-5	-10	24	2
Total BHF	389	13,703						78

*Reflects typical rate after water is diverted into facility; lower rates may prevail within long periods between inflow events.

Abbreviations: ft. MSL= feet mean sea level; ac-ft = acre-feet; ac-ft/d= acre feet per day.

Pits complex, include Kaiser-BHF Pond, Shinn Pond, and Pits T-1 and T-2. Water to Kaiser Pond-BHF is received from Kaiser-AHF Pond via a pipe through the levee that separates these two ponds. Pit T-1 is supplied directly from Alameda Creek water impounded behind Rubber Dam No. 2, and Pit T-2 is supplied from Pit T-1. Diversions to the Quarry Pits occur through diversion pipes that convey water from Shinn Pond and empty into Stevenson Pond and Rock Pond. These two ponds, in turn, supply the rest of the Quarry Pits.

In the summer of 1997, several months before the study, the Quarry Pits complex underwent significant re-grading at elevations above 0 feet MSL. Steep pond side-slopes, resembling bluffs, were cut back to gentler grades to improve structural stability, reduce erosion, and to enhance accessibility to the water line. In addition, the tops of levees were removed in order to create large ponds from former smaller ponds. Lago Los Osos was constructed by joining former Pits R-1, R-2, R-3 and S. Similarly, Horseshoe Pond was created from former Pits G and H. The levees were not excavated to their base elevations; therefore, the bottom of Lago Los Osos and Horseshoe Pond have irregular bathymetry, with multiple deep sections located near the centers of the former smaller ponds that were joined to form these two subject larger ponds. The smaller ponds, in effect, are still preserved below 0 feet MSL. The irregular bathymetry is of interest because the potential for resistance to horizontal mixing.

Clogging Layers and Seepage Rates

Deposition of low-permeable sediment is believed to have resulted in the formation of clogging layers that tend to slow rates of seepage from ACWD's recharge ponds and, possibly, Alameda Creek. Previous field observations and analysis undertaken by ACWD have provided evidence of the existence of clogging material. Permeability tests on sediment samples collected from Shinn Pit in 1963 indicated a permeability value of 10^{-5} feet/day in the first one inch of sediment collected from the side slope. Permeability increased with penetration distance behind the slope, with values of 1 foot/day for a sample collected 1 to 2 feet of penetration distance, and in excess of 10 feet at penetration distances greater than 2 feet. In April 1979, ACWD investigated options for increasing percolation in Kaiser Pit in response to seepage rates having dropped from 10 cfs in 1971 to 3 cfs at the time of the (Banks 1979) study. Water levels were lowered from their typical operating range of 50+ feet MSL to an elevation 33 feet MSL. Approximately 700 linear feet along the perimeter of the northwest part of the pit was scarified between elevations 30 and 50 feet MSL. Subsequently, after pit levels were returned to their normal operating range (50 to 52 feet MSL) in June 1979, percolation increased to 20 cfs, but dropped back to 10 cfs by September 1979. The consultant's report (Banks 1979) concluded that a combination of scarification and drying were responsible for the observed increase in percolation. ACWD also observed notable increases in percolation rates in the Quarry Pits following their reconstruction in 1997. To accommodate this earthwork, water levels in the pits were lowered from their normal operating range (which, prior to this project were 10 to 28 feet MSL) to –3 feet MSL. Formerly submerged clogging materials were inevitably scarified during re-grading or were allowed to dry. After completion of the project, seepage rates were approximately 30 percent higher than they were prior to the project (unpublished data, ACWD 1998). However, ACWD has not determined the extent to which scarification and drying were responsible for this increase. Higher operating levels and re-shaping of the ponds probably also contributed to the improved percolation rates.

Clogging material may have been formed or deposited in a variety of ways, including settling of inorganic and/or organic particles carried by waters diverted into the ponds, erosion of top soil from the steep pond side slopes, and biological growth at the soil/water column interface. Quarry operators' disposal of fine-grained material washed from the excavated gravel, before ACWD converted the ponds for artificial recharge operations, may also be responsible for some of the bottom sediment. Aside from occasional earthwork projects described above, there is no regular maintenance program to remove clogging materials from ACWD's recharge facilities.

The influence of pond sediment on recharge rates was investigated to help interpret the results of the tracer experiments, and the results are discussed in Chapter 4. In addition to investigating seepage from ponds, the study has also attempted to examine the effect that pond sediments exert on the quality of water reaching the production wells.

Sources of Recharge and Potable Groundwater

The natural baseflow in Alameda Creek is derived from drainage of the 600 square-mile watershed up-stream of ACWD (Figure 2.1). In the past few years, groundwater pumped from gravel quarries (not part of ACWD's recharge operation, but located in the upstream watershed) into the creek has been a significant component of this base flow, especially in the dry summer months. Additional water is imported from the California State Water Project when the amount of creek water supplied from local watershed sources falls short of ACWD's recharge requirements. State water delivered to ACWD is derived largely from snowmelt in the Sierra Nevada Mountain range along the California-Nevada border. A quantitative breakdown of sources of surface water percolated as artificial recharge is provided in Chapter 4.

Although water from the recharge facilities makes up a large portion of the supply feeding the production wells, the wells may also be fed in part by groundwater derived from local interflow from the hills or percolation from direct rain. The study included an investigation of the dominant source of water in the various production wells.

Production Wells and Monitoring Wells

The locations of individual monitoring wells and production wells are provided in Figures 2.4 and 2.8, respectively, and well construction data are included in Tables 2.3 and 2.4.

AHF Wells

The PT Wellfield is comprised of eight production wells, PT 1 through 8, which are arranged in a general south-to-north alignment and spaced approximately 100 feet apart (Figure 2.8). The northernmost wells, PT-7 and PT-8, are closest to Kaiser Pit, with a pond-to-well distance of approximately 1,200 feet. In contrast, the southernmost well, PT-1, is approximately 1,800 feet from Kaiser Pond. Bunting Pond is situated approximately 1,800 feet east-northeast of the general alignment of PT wells. The total depth of each well in the PT Wellfield is approximately 200 feet, which is estimated to be about halfway to the bottom of the aquifer. The discharge rate of each of the eight PT wells is typically 2,200 gpm (10 acre-feet/day) (Table 2.4). Wells are individually metered for discharge, and ACWD maintains daily production records on a well-by-well basis.

In the spring of 1997, several months before dosing, ACWD installed monitoring well 21L03 located halfway between Kaiser Pond and the Peralta-Tyson Wellfield (Figure 2.4). Because the depths of the fastest streamlines in the aquifer were unknown at the time, the well was constructed to intercept a large thickness of the aquifer. The intent of this well was to flag the presence or absence of tracer in the aquifer, regardless of the elevation of any pathway the tracer may have followed. The well has a total depth of 200 feet below ground surface, and has

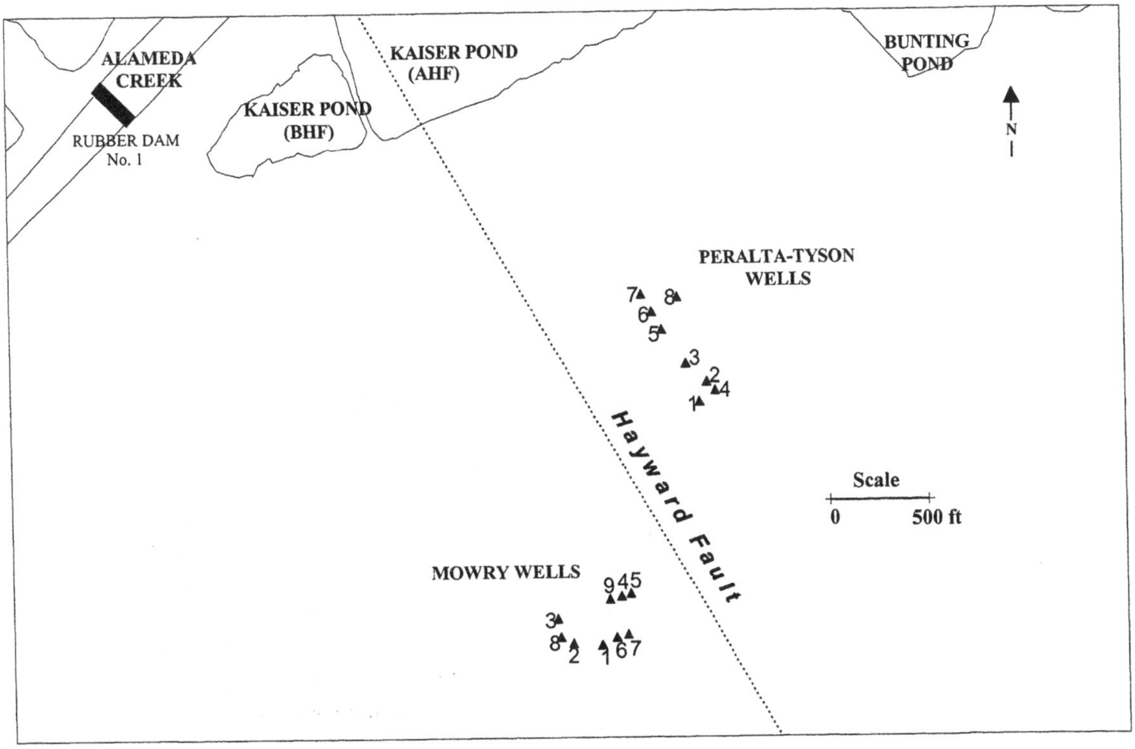

Figure 2.8 Peralta-Tyson (PT) and Mowry production wells, located in AHF and BHF portions of the study area, respectively

a single casing with three screened intervals: a shallow zone, a middle zone, and a deep zone (see Table 2.3 and Figure 2.6). These three intervals are separated from each other only by sections of blank casing, and there is no packer to separate the three zones. Consequently, vertical mixing is possible inside the casing and in the filter pack outside the casing.

In November and December of 1997, ACWD converted five one-inch diameter piezometers, located immediately adjacent to the southern shorelines of Kaiser and Bunting Ponds, into two-inch monitoring wells. Three of the wells, 21L05, 21L06, 21L07, were used to monitor tracer seeping from Kaiser Pond, while tracer from Bunting Pond was monitored by wells 21J03 and 21K05 (Figure 2.4). The total depths of these five converted wells range from 70 to 100 feet. Other monitoring wells include 21L08, which is the closest well to Kaiser Pit, and 21J27. Well 27D08, an outlying well on the edge of the study area, was monitored to help define the extent of southerly travel of the tracer in the AHF sub-basin.

To provide the opportunity to monitor groundwater impacts from Alameda Creek upstream of the recharge ponds, ACWD converted 21H11 from a 1-inch piezometer to a 2-inch

Table 2.3
Monitoring well data

Well	Sub-basin	Approx. ground eleva. (ft. MSL)	Total depth (ft. b.g.s.)	Casing diam. (in)	Casing material	Screened interval top (ft. b.g.s)	Screened interval bottom (ft. b.g.s.)	Aquifer penetrated by screen
20J04	BHF	58	57	2	PVC	30	50	Newark
20J05	BHF	58	110	4	PVC	99	109	Newark
20J06	BHF	58	80	2	PVC	70	80	Newark
20R02	BHF	65	150	8	steel	N.A.	N.A.	Newark
20R03	BHF	58	60	2	PVC	32	60	Newark
20R04	BHF	58	85	2	PVC	65	80	Newark
20R05	BHF	58	115	2	PVC	95	115	Newark
20R07	BHF	63	75	2	PVC	45	75	Newark
21H11	AHF	76	75	2	PVC	30	70	AHF
21J03	AHF	77	90	2	PVC	40	80	AHF
21K05	AHF	70	94	2	PVC	40	80	AHF
21L03	AHF	67	202	4	PVC	70	90	AHF
						125	145	
						175	195	
21L05	AHF	67	70	2	PVC	48	68	AHF
21L06	AHF	68	70	2	PVC	46	66	AHF
21L07	AHF	71	70	2	PVC	46	66	AHF
21L08	AHF	67	99	2	PVC	30	60	AHF
27D08	AHF	66	110	8	PVC	60	110	AHF
28D01	BHF	N.A.	457	4	PVC	415	445	Deep
28D08	BHF	N.A.	385	4	PVC	370	380	Fremont
28D11	BHF	63	290	4	PVC	240	280	Centerville
28D12	AHF	63	165	4	PVC	120	160	Newark
21J27	AHF	74	70	4	PVC	50	70	AHF

Abbreviations: b.g.s. = below ground surface; MSL= mean sea level, N.A.= not available.

monitoring well in April 1998. This well is located 570 feet northwest of Bunting Pond and 250 southeast of Alameda Creek.

BHF Wells

The Mowry Wellfield is comprised of nine wells (Figure 2.8), six of which were active during the study period for this project. Mowry Wells 1 and 7 extract water from the Newark Aquifer; Mowry Wells 3, 6, and 9 are screened in the Centerville Aquifer and Mowry 2 is perforated in the Fremont and the Deep Aquifer. Like production wells in the Peralta-Tyson

Table 2.4

Production well data

Well	Status during project	Typical discharge rate (ac-ft/d)	Approx. ground elevation (ft MSL)	Total depth (ft b.g.s.)	Casing diameter (in)	Casing material	Perforated interval* top (ft b.g.s)	Perforated interval* bottom (ft b.g.s.)	Aquifer penetrated by perforated interval
Above Hayward Fault									
PT-1	active	10	65	200	18	steel	68	185	AHF
PT-2	active	10	65	200	18	steel	75	190	AHF
PT-3	active	10	66	200	18	steel	80	177	AHF
PT-4	active	10	66	200	18	steel	73	180	AHF
PT-5	active	10	66	200	18	steel	57	184	AHF
PT-6	active	10	66	200	18	steel	60	184	AHF
PT-7	active	10	66	200	18	steel	54	186	AHF
PT-8	active	10	67	215	18	steel	120	215	AHF
Below Hayward Fault									
Mowry 1	active	3	62	194	12	steel	122	182	Newark
Mowry 2	active	10	61	470	18	steel	345	455	Fremont, Deep
Mowry 3	active	10	62	311	16	steel	197	305	Centerville
Mowry 4	inactive	N.A.	63	196	18	steel	70	180	Newark
Mowry 5	inactive	N.A.	63	332	16	steel	226	320	Centerville
Mowry 6	active	10	62	310	18	steel	220	300	Centerville
Mowry 7	active	10	62	190	18	steel	120	180	Newark
Mowry 8	inactive	N.A.	61	190	18	steel	120	180	Newark
Mowry 9	active	10	63	320	18	steel	220	300	Centerville

* Some sections of blank casing may exist within the indicated interval.

Abbreviations: ac-ft/d= acre-feet per day; ft b.g.s = feet below ground surface; ft MSL= feet mean sea level; N.A.= not applicable.

Wellfield, Mowry wells are individually metered and ACWD maintains daily production records for each well. The production rate from each Mowry well is typically 2,500 gpm (11 acre-feet/day), except Mowry 1 which generally discharges at a rate of 600 gpm (2.7 acre-feet/day).

Between the Mowry Wellfield and the BHF recharge ponds, groundwater was sampled from up to twelve wells grouped in four clusters (Figure 2.4). Two of the clusters are situated close to Alameda Creek, and each consists of three wells to monitor the upper, middle, and lower depths, respectively, of the Newark Aquifer. The six wells that comprise these two clusters are 20J04, 20J05, 20J06, 20R03, 20R04 and 20R05. Wells 28D01, 28D08, 28D11 and 28D12 comprise the third of the four aforementioned clusters. These four wells, located approximately 1,400 feet northwest of the Mowry Wellfield, are screened at depths representative of the Deep, Fremont, Centerville, and Newark Aquifers, respectively. ACWD added two wells to the BHF monitoring well network to comprise the fourth and final cluster. The two wells, 20R02 and

20R07, are laterally separated from each other by only 140 feet, but have different screened intervals within the Newark Aquifer (Table 2.3).

Well 20R02, destroyed shortly after completion of field activities for this study, was a privately-owned domestic/irrigation well equipped with its own dedicated pump. With a total depth of 150 feet and with the elevation of the top of the perforations unknown, the well may have extracted water from the lower half of the saturated thickness of the Newark Aquifer. Because of the concern that the tracer might travel in groundwater preferentially at shallow depths, monitoring well 20R07 was constructed in June 1998. It has a relatively shallow total depth of 75 feet, and its screened interval is between 45 and 75 feet below ground surface.

CHAPTER 3

METHODS AND MATERIALS

ELEVATIONS OF PIEZOMETRIC HEAD

An electronic probe, accurate to within 0.01 foot, was utilized for measuring the depth to water in each well from the top of casing. Elevation of piezometric head was calculated by subtracting the depth to water from the top-of-casing elevations. At wells 20J04, 20J05, 20J06, 20R03, 20R04, and 20R05, the top-of-casing elevations were measured by conventional surveying shortly after these wells were constructed in the mid-1990s. The top-of-casing elevations at the remaining wells were measured by a highly accurate global positioning system (GPS) (Trimble Navigation Instruments Model 4400, Sunnyvale, California) that became available at the end of the study period. Unfortunately, the elevations of the 20J04, 20J05, 20J06, 20R03, 20R04, and 20R05 could not be checked with the GPS unit because the well heads were physically altered during land development activities. Disturbance of the wells occurred during the period following the final round of water level measurements but prior to the availability of the GPS system. However, their horizontal position was not affected, and the GPS unit was used to survey the location (latitude and longitude) of every well in the study.

KAISER POND SEEPAGE CALCULATION

Diversion pipes that convey water from Alameda Creek into recharge ponds are equipped with either flow rate meters or totalizer meters (totalizers record cumulative volume of water that has passed through the pipe, not instantaneous flow rate). Generally, the amount of water flowing in the ponds is recorded by ACWD on a daily basis. In addition, ACWD maintains daily records of elevations of surface water impounded by each of the three rubber dams in the creek as well as water level elevations at most of the recharge ponds. These elevations are read directly by observation of marked steel pipes anchored in the bottoms of the ponds and creek segments so equipped. The daily records include the hour and minute of the day of each measurement.

Some of these data were entered in a mass balance algorithm to calculate rates at which water seeped from Kaiser Pond. For a given time increment Δt,

$$\textit{Change in storage} = (\textit{volume in}) - (\textit{volume out}). \tag{3.1}$$

Rearranging,

$$\textit{volume out} = \textit{volume in} - \textit{change in storage} \tag{3.2}$$

The inflow volume was calculated from inflow pipe totalizer readings, and the change in storage was determined by noting the observed change in pond level and against a pre-determined relationship between stage and storage. The seepage rate for the time increment was then calculated by dividing the volume out by Δt:

$$\textit{seepage rate} = (\textit{volume out})/\Delta t \tag{3.3}$$

The relationship between seepage rate and various parameters, such as pond level and inflow rate, is discussed in Chapter 4. Inflow rate was taken to be:

$$\textit{inflow rate} = (\textit{volume in}) / \Delta t \tag{3.4}$$

For calculation of recharge rates and inflow rates, each Δt was calculated using the time data (minute and hour) coinciding with a reading of the totalizer meter and the recording of the pond level (pond levels and meter readings were recorded at the same time). These time increments worked out to be approximately two days and were centered about a "date of interest". Dates of interest coincided with groundwater level measurements considered for comparison in this analysis. The beginning of each time step coincided with the time the flow meter reading and pond level were recorded on the day before the date of interest. Similarly, the end of the time step coincided with a flow meter and pond level reading on the day after the date of interest. Well water levels were recorded typically on a weekly basis between January 1997 and June 1999. For this analysis, 48 of 76 groundwater level measurements at well 21L08

(AHF) and 26 of 40 at 20J04 (BHF). The remaining measurements could not be considered due to gaps in pond data on the required dates. Since pond levels were recorded daily (except when the totalizer meter was not working or when pond levels were below the minimum staff level), a pit water level measurement was available on each date of interest and on the days coinciding with the beginning and end of each time step. For calculation of correlation coefficients, the recharge rate was plotted as a function of the following parameters: groundwater level, pond water level, and the difference between pond level and groundwater level. Finally, a correlation coefficient was calculated for the relationship between recharge rate and inflow rate.

INITIAL STUDY TO DETERMINE ABUNDANCE OF NATURAL ISOTOPES AND NOBLE GASES

The content of selected natural isotopes and dissolved noble gases in ACWD surface water and groundwater was measured in the spring 1997. Targeted isotopes included deuterium, tritium, Helium-3, Helium-4, Oxygen-18, Carbon-13, and Carbon-14. Noble gases selected for analysis included xenon (specifically, the dominant isotope of xenon, Xenon-132), krypton, neon, and argon. Natural isotopes were analyzed to determine average groundwater ages in selected wells, to differentiate among surface water sources of groundwater in the wells, and to help explain processes that are responsible for the chemistry of ACWD's groundwater. The concentrations of dissolved noble gases were determined for estimating the temperature of groundwater at the time that it entered the aquifer.

The methodology by which the abundances of isotopes were interpreted for age dating in this study is provided in Hudson et al. (1995). Hudson et al. (1995) and Hackley et al. (1996) provide the basis for the application of isotopic methods for geochemical interpretation and differentiation of recharge sources. Historical groundwater temperatures were estimated based on methodology described in Aeschbach-Hertig et al. (1999).

Determination of Historical Groundwater Temperature

Equilibrium concentrations of noble gases in surface water depend on temperature, salinity, and ambient atmospheric pressure. Because the solubility curve is distinct for each noble gas, a robust estimate of the temperature of recharge can be made based on measurements of the dissolved abundance of He, Ne, Ar, Kr, and Xe. After percolation, the concentration of these dissolved gases in groundwater is preserved, enabling a determination of temperature at the time of recharge to the aquifer. The methodology for estimating historical groundwater temperature in this study is further explained in a presentation of the results in Chapter 4.

Age Dating by the Tritium-Helium Method

A short-lived radioactive isotope of hydrogen (^3H), called tritium, occurs in the atmosphere and hydrosphere as a small component of the water molecule. After water in isotopic equilibrium with the atmosphere is recharged to groundwater, the tritium isotope decays (half life is 12.4 years), enabling a determination of the average age of the groundwater, measured from time of recharge:

$$\text{age} = -17.9 \times \ln\left(\frac{^3H}{^3H_0}\right) \tag{3.5}$$

where ^3H= tritium content measured in the sample

and ^3H_o= the tritium content in the water at the time of recharge.

Atmospheric testing of thermonuclear weapons in the 1950s and 1960s increased tritium levels in precipitation and surface water throughout the world by several orders of magnitude. Tritium fallout to the earth's surface depends upon latitude and precipitation patterns. Hence ^3H_o is difficult to estimate. However, the amount of tritium that has actually decayed (and thus the ratio of ^3H to $^3H_0$) can be calculated with a $^3He_{trit}$, which is the component of ^3He derived from the decay of ^3H. Since $^3H_o = {}^3H + {}^3He_{trit}$, the above equation can be re-written as:

$$\text{age} = -17.9 \times \ln\left(\frac{^3H}{^3He_{trit} + {}^3H}\right) \tag{3.6}$$

$^3He_{trit}$ is only one of multiple components of the amount of ^3He measured in water, as indicated by the following expression:

$$^3He_{meas} = {}^3He_{trit} + {}^3He_{equil} + {}^3He_{excess} + {}^3He_{rad} \quad (3.7)$$

where $^3He_{meas}$ = the measured total amount of ^3He in the sample,

$^3He_{trit}$ = the amount of ^3He due to decay of ^3H,

$^3He_{equil}$ = the amount of ^3He dissolved in non-turbulent water in equilibrium with the atmosphere,

$^3He_{excess}$ = the amount of ^3He dissolved in water attributed to excess dissolved air and

$^3He_{rad}$ = the amount of ^3He due to radioactive decay from isotopes other than tritium.

The amount of $^3He_{rad}$ can be ignored since it makes up, at most, only 0.2% of $^3He_{meas}$ (Schlosser et al. 1989).

The ratio that is measured by noble gas mass spectrometry in the laboratory will be denoted herein as $(^3He/^4He)_{water}$ to distinguish the $^3He/^4He$ ratio in water from that in air. Ratios for both media are referred to in this section. Measurement of the ^4He content can also help distinguish the components of ^3He present due to interaction with the atmosphere (when air bubbles are entrained as water moves through the vadose zone) or 'excess air' in the water ($^3He_{equil}$ + $^3He_{excess}$), and hence, enable a back-calculation of $^3He_{trit}$. The determination is possible because the ratio of atmospheric or air-derived ^3He and ^4He are known. Hence:

$$^3He_{equil} + {}^3He_{excess} \approx \left(\frac{^3He}{^4He}\right)_{air} \times {}^4He_{meas} \quad (3.8)$$

$^3He_{trit}$ can then be calculated as:

$$^3He_{trit} \approx {}^3He_{meas} - ({}^3He_{equil} + {}^3He_{excess}) \quad (3.9)$$

When groundwater age is greater than 1,000 years old, ^4He above atmospheric levels could be present due to uranium-thorium decay (Schlosser et al. 1988), and measurement of ^4He

will not enable reliable determination of $^3He_{equil}$ + $^3He_{excess}$. To overcome this problem in older waters, other atmospheric components such as neon, argon, and krypton can be measured to estimate the amount of excess air in the water, and hence, $^3He_{equil}$ + $^3He_{excess}$. In order to give as much confidence as possible to the estimate of $^3He_{equil}$ + $^3He_{excess}$, it's advisable to measure these additional elements even in waters anticipated to be relatively young.

Because groundwater moves slowly, ages can vary substantially over short distances. However, over small areas (such as the study area for this project) within a single aquifer layer, there's typically not much variation in $^3H_0$ or the amount of excess air. Hence, after establishing representative values for these two parameters from a sufficient number of samples, subsequent samples for age dating collected over a limited area need only to be analyzed for ^3He (that is, $^3He_{meas}$). The value of $^3He_{trit}$ can then be inferred by subtracting the assumed value of ($^3He_{equil}$ + $^3He_{excess}$) from $^3He_{meas}$. If ^4He has also been pre-determined to not vary significantly with the targeted study area, samples need only be measured for $(^3He/^4He)_{water}$. Ages are then calculated by:

$$\text{age} = -17.9 \times \ln\left(\frac{^3H}{^3H_0}\right) = -17.9 \times \ln\left(\frac{^3H_0 - {}^3He_{trit}}{^3H_0}\right) \qquad (3.10)$$

Although theses shortcuts can somewhat reduce the accuracy of the age values, they can substantially cut down on laboratory analytical costs when a large number of samples need to be collected over a limited area. This approach was followed for a few samples collected during this project.

Because water in a well is drawn in from all radial directions, the water column in a well, particularly a production well, is a mixture of water particles that likely have a range of ages. The 'water age', determined using the tritium-helium method, is the mean age of the tritium in the water sample. The groundwater ages reported herein (Chapter 4) are mean, integrated values.

Stable Isotopes of Oxygen and Hydrogen to Distinguish Sources of Recharge

The dominant isotopes of hydrogen and oxygen that overwhelmingly account for most of the water molecules in the environment are, respectively, ^1H and ^{16}O. However, a small portion of the water molecules will consist of ^2H (commonly referred to as deuterium, denoted as *D*) and/or ^{18}O. The lighter isotopes are enriched in water vapor formed by evaporation of liquid water, while the remaining fraction is enriched in ^{18}O and *D*. Rain falling from clouds that have moved from coastal areas to inland (colder, higher elevation) regions becomes progressively enriched in the lighter isotopes (negative $\delta^{18}O$). The abundance of these isotopes in groundwater samples can provide clues as to the origin of the source water from which the groundwater is derived. Because the molar concentration of these non-dominant isotopes are very low, the so-called "del" notation is used as a more convenient measure of their content in a water sample. It provides an indication of the abundance of the target isotope in the sample relative to the dominant isotope of the same element (also present in the sample). This ratio, in turn, is expressed relative to the isotope ratio of a standard. Subtracting from 1 and multiplying by 1,000 completes the calculation and gives a useful measure of the deviation from the standard; i.e.,

$$\delta D = \left[\frac{(D/^1H)_{sample}}{(D/^1H)_{standard}} - 1 \right] \times 1000 \qquad (3.11)$$

and

$$\delta^{18}O = \left[\frac{(^{18}O/^{16}O)_{sample}}{(^{18}O/^{16}O)_{standard}} - 1 \right] \times 1000 \qquad (3.12)$$

The units of the del convention are "per mil," with the symbol $^{o}/_{oo}$. Positive and negative del values indicate, respectively, a degree of enrichment and depletion of the isotope relative to the standard.

The $(D/^1H)$ standard and $(^{18}/^{16}O)$ are the isotopic ratios of the NBS-standard "Standard Mean Ocean Water" (SMOW). The values of δD and $\delta^{18}O$ in a water sample will tend to closely follow a straight-line relationship known as the "Global Mean Meteoric Water Line" (MWL), represented by the following equation:

$$\delta D = 8 \times \delta^{18}O + 10 \qquad (3.13)$$

The equation was developed by regression of the δD and $\delta^{18}O$ values of numerous samples of natural waters throughout the world (Craig 1961). A plot of the line is provided in Chapter 4. Where exactly on the line a particular water sample will plot will depend on geographical location and elevation of the ground surface on which the precipitation falls. Water samples from lower latitudes will plot lower on the line and vice-versa. In California, $\delta^{18}O$ in precipitation varies on average from approximately -5 $^{o}/_{oo}$ along the Pacific coast to -14 $^{o}/_{oo}$ in the Sierra Nevada Mountains.

After percolation to groundwater, the δD and $\delta^{18}O$ signature will generally be preserved (except in geothermal areas) and, consequently, can provide an indication of the surface water source of the target groundwater (Criss 1999). For example, groundwater near recharge ponds, supplied largely by mountain ranges located many miles inland of the groundwater basin, should plot lower on the MWL compared to groundwater in the basin derived from local rainfall.

Certain physical and chemical processes in certain waters can change the D and ^{18}O content and will cause the position of their δD-$\delta^{18}O$ plot to move off the MWL to some extent. Molecules of surface water consisting of the lighter hydrogen (^{1}H) and oxygen (^{16}O) isotopes will preferentially be evaporated over those comprised of D and/or ^{18}O , moving the δD-$\delta^{18}O$ point upward along the MWL but at a lesser slope than the MWL (and therefore will plot below the MWL line). Other physical, chemical, or biological processes occurring in surface water or groundwater can shift the δD-$\delta^{18}O$ signature off the MWL (Criss 1999). They include carbon dioxide exchange, hydrogen sulfide exchange reactions, fluctuations in temperature, water/rock interactions, hydration of silicates, and methanogenesis.

Analysis of Carbon for Characterizing Age and Water Chemistry

The dominant carbon isotope in nature is Carbon-12 (^{12}C). The less abundant heavier isotopes are ^{13}C and ^{14}C, and, like D and ^{18}O discussed above, their content relative to the abundant isotope can also be expressed in the del notation; i.e.,

$$\delta^{13}C = \left[\frac{(^{13}C/^{12}C)_{sample}}{(^{13}C/^{12}C)_{standard}} - 1 \right] \times 1000 \qquad (3.14)$$

Nuclear weapons testing in the 1950's and 1960's increased the ^{14}C content in the upper atmosphere (the ^{14}C content in the atmosphere remains slightly elevated at the present time). Samples with such excess ^{14}C are said to have a "bomb-pulse" ^{14}C signature. The ^{14}C content can be expressed as "percent modern carbon" (pmC), where 100 pmC corresponds to "natural" ^{14}C abundance (relative to ^{12}C) in the atmosphere prior to nuclear weapons testing.

In groundwater, $\delta^{13}C$ and $^{14}C/C$ measurements can be interpreted in various ways to characterize groundwater, since groundwater contains organic carbon (i.e. TOC) and dissolved inorganic carbon (DIC) species (CO_2, H_2CO_3, HCO_3^-, and CO_3^{-2}). For this project, carbon isotopes were measured from DIC species only, not TOC. The inorganic carbon may be derived from inorganic reactions, such as dissolution of minerals, or from microbially-mediated mineralization of dead organic matter.

Carbon dioxide generated from microbial aerobic oxidation of terrestrial plants will have a $\delta^{13}C$ of -10 to -35 $^{o}/_{oo}$, whereas the $\delta^{13}C$ in the component of DIC attributed to the dissolution of calcium carbonate is 0 $^{o}/_{oo}$. As is the case with aerobic decomposition, CO_2 produced from anaerobic microbial oxidation of dead organic matter (e.g., denitrification and sulfate reduction) will tend to be enriched in ^{12}C. Consequently, the $\delta^{13}C$ in organic mater left behind will be increased. Not all microbial reactions, however, will decrease the $\delta^{13}C$ in CO_2. For example, methanogenic bacteria will preferentially utilize the ^{12}C in CO_2 to produce ^{12}C-enriched methane, and this will increase $\delta^{13}C$ in the remaining of CO_2 (Hoefs 1980).

Because ^{14}C decays slowly to ^{12}C (half-life is 5,730 years), measurements of ^{14}C content can be used to detect very old groundwater (over 400 years old) or, at least, very old carbon from minerals dissolved in groundwater (Wigley 1975). Groundwater of such age would be expected to have less than 100 pmC. Also, younger groundwater whose DIC content is heavily owed to dissolution of very old carbonate minerals would also be expected to have less than 100 pmC.

The ^{14}C content of plants grown since the beginning of atomospheric nuclear weapons testing will be enriched in ^{14}C commensurately with atmospheric bomb-pulse ^{14}C. Consequently, water percolated to the subsurface after 1960 will have greater than 100 pmC if its DIC content is derived primarily from either (1) dissolution of atmospheric CO_2 (the dissolution occuring, of course, while the water was still surface water prior to percolation as groundwater), or (2) soil gas (which would occur mainly at the time of percolation through the vadose zone) whose CO_2 content is derived from respiration of plants grown after 1960. But as indicated above, interaction with very old carbonate minerals could bring the ^{14}C abundance in young groundwater back down to, or below, 100 pmC.

The initial study for ACWD was as conducted in the spring of 1997. Groundwater samples were collected from production wells in both wellfields and from monitoring wells in the AHF sub-basin for analysis of $\delta^{18}O$, δD, $\delta^{13}C$, ^{3}H, ^{3}He, and ^{4}He, Ne, Ar, and Kr, and ^{14}C.

TRACER STUDIES

Tracer studies were needed because groundwater age estimates alone were not sufficient to determine the travel times of water from recharge ponds to the wellfields. Moreover, tracers enable determination of preferential pathways.

AHF Recharge Pond Tracer Experiment

The AHF pond tracer experiment commenced with the introduction of ^{136}Xe and ^{124}Xe tracers to Kaiser (AHF) and Bunting Ponds, respectively. Tracer was loaded in four one-day visits, each spaced six to seven days apart, beginning on December 16, 1997 and ending on January 6, 1998. This gradual loading was intended to ensure the resulting spike of tracer moving through groundwater would not be too short in duration, which would require an unreasonably high monitoring frequency. Application of a unique tracer for each pond was intended to enable determination of the travel time of water from each individual pond to the production wells, instead of a mere estimate of travel time for the two recharge ponds as a group. Like other tracers used in this project, these two xenon tracers provided a chemically inert, non-toxic and non-radioactive tracer with a large dynamic dilution range. Both the California Department of Health Services and the California Regional Water Quality Control Board were contacted in advance of the experiment, and each agency responded positively without imposing special restrictions.

Each tracer was prepared in the laboratory (Lawrence Livermore National Laboratory) by pre-dissolving the tracer gas in water so that it could be conveniently gravity fed as a solution through a hose to the desired pond depth. Tracer was added to each pond from a single station near the approximate center of each pond. Access was provided by a small boat, which accommodated two persons plus the dosing equipment. Before tracer was added, a temperature profile of each pond was developed to determine if a thermocline was present. As no thermocline was evident in either pond, the tracers were introduced at a depth of 25 feet.

The concentration of tracer in the ponds was measured on a weekly basis throughout the dosing period and for two weeks after the final dosing event. Sampling and dosing were typically performed within the same trip to the ponds, with samples being collected just before additional tracer was added (no samples were collected on the first day of dosing). To check for possible vertical stratification and incomplete horizontal mixing, samples were collected from at least two stations in each pond and from two to three depths at each station. Although vertical stratification was not anticipated due to the apparent absence of a thermocline, samples were collected from both deep and shallow depths.

Locations within Kaiser and Bunting Ponds at which tracer was added or samples collected are indicated in Figure 3.1.

Well monitoring to determine the response of groundwater to the introduction of tracers to the AHF ponds commenced on January 6, 1998, coinciding with the last day of tracer introduction to the AHF ponds. Samples were collected from selected wells on a weekly to biweekly basis through July 1999. A final set of samples was collected in December of 1999. The first wells sampled were the monitoring wells immediately adjacent to the ponds. Gradually, more down-gradient monitoring wells, and finally the production wells themselves, were included in the monitoring program.

In addition to collecting samples for tracer content, samples for characterization of natural pond water chemistry were collected on December 29, 1997, coinciding with the third week of dosing (see "Water Quality Characterization during Tracer Studies" later in this chapter). Samples were collected from two stations in each pond and from two depths per station, giving a total of four samples per pond. The depths from which samples were collected were approximately three feet below the surface and three feet above the bottom.

BHF Pond Tracer Experiment

Neon-22 was added to two BHF ponds, Stevenson Pond and the Pit S portion of Lago los Osos, on August 17, 20, 25 and 27, 1998 (Figure 3.2). Of all ponds in the Quarry Pits complex, these two ponds were thought to be the most likely to recharge water that would reach the Mowry Wellfield within the time frame of the study. Dosing of Stevenson Pond was achieved

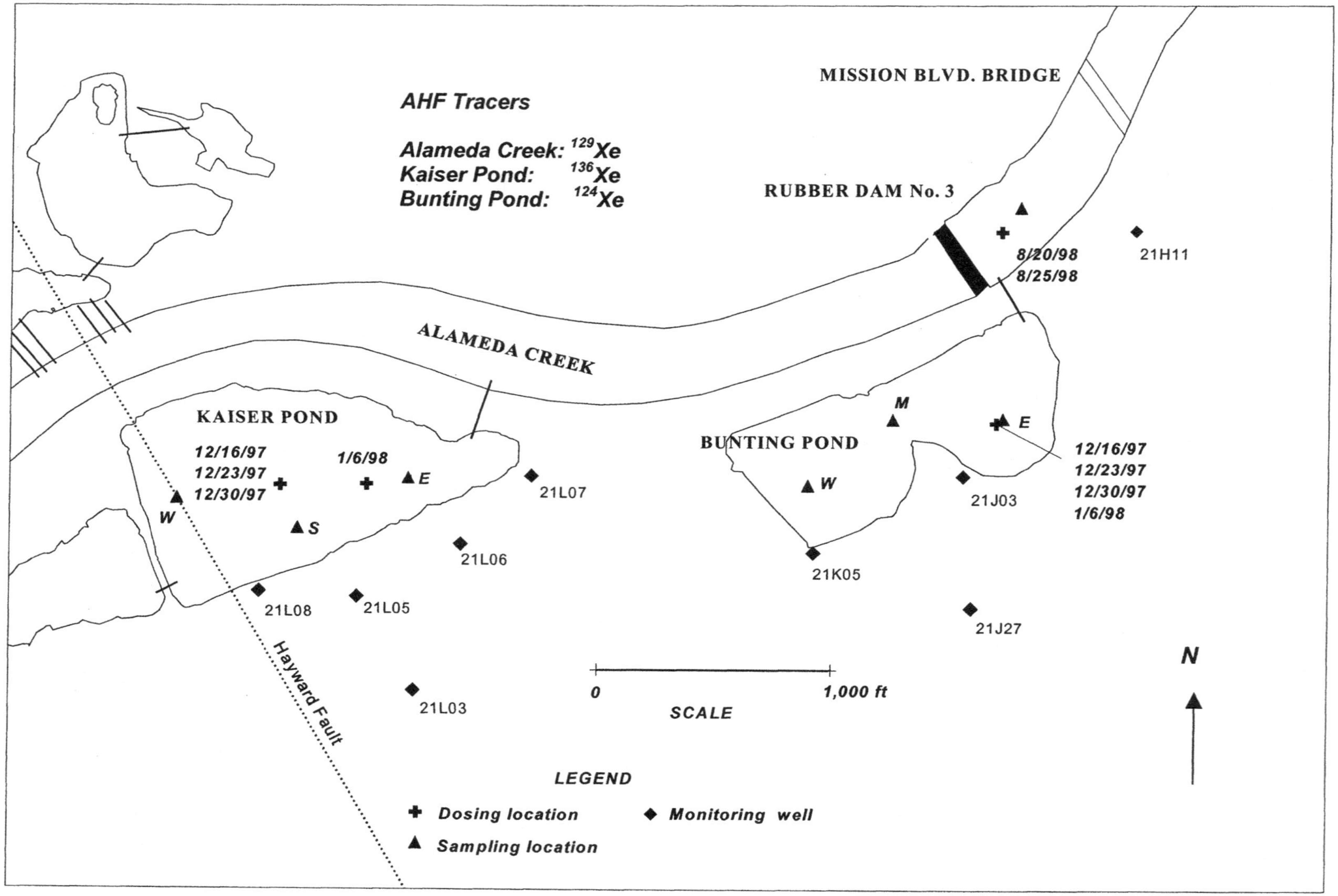

Figure 3.1 Dosing and sampling locations for AHF tracer experiments. Dates on which dosing occurred are included next to the dosing station. Pond sampling stations are denoted as E (east), M (middle), S (south), or W (west).

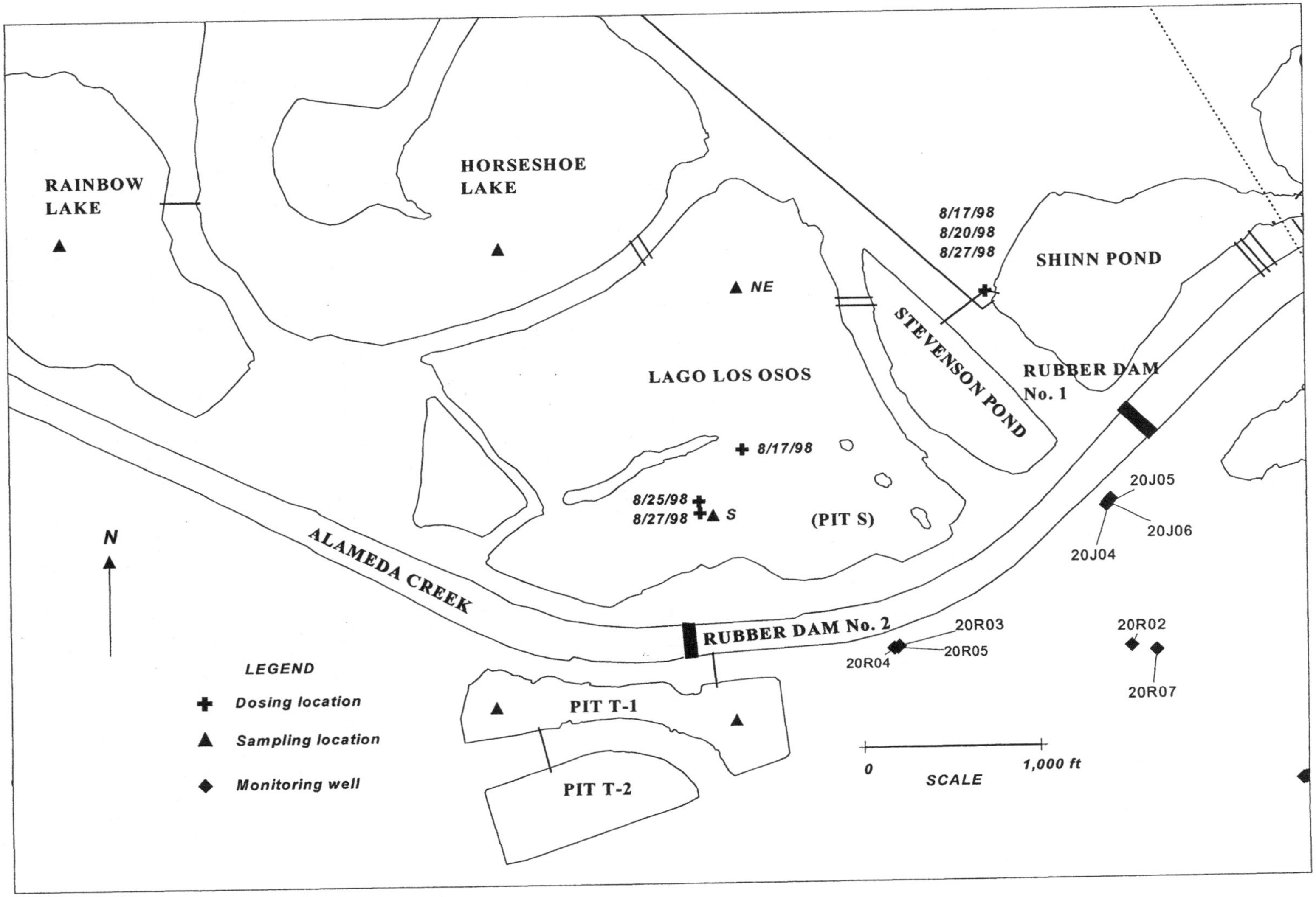

Figure 3.2 Dosing and sampling locations for BHF tracer experiment (^{22}Ne). Dates on which dosing occurred are included next to the dosing station. Lago los Osos sampling stations are denoted as NE (northeast) and S (south).

by introducing tracer into a manhole of a diversion pipe that carries water from Shinn Pond to Stevenson Pond. By opening a valve, a relatively small amount of water was diverted from Shinn Pond to Stevenson Pond in order to wash the tracer through the pipe. Some of the ^{22}Ne added to Stevenson Pond was inevitably transferred into Pit S because the two ponds are interconnected by a gravity-flow pipe that has no valve. To minimize this transfer, only a small amount of water was diverted into Stevenson Pond, thereby preventing the water level in Stevenson Pond from rising significantly higher than that of Pit S. It was decided not to load tracer into Pit S by a heavy through-put of water from Shinn Pond to Pit S via Stevenson Pond: the introduction of such a substantial quantity of new water in the pits would have diluted the tracer, requiring more tracer to be added. Instead, most of the tracer added to Pit S was directly loaded by boat according to the same procedure followed for the earlier dosing of Kaiser (AHF) and Bunting Ponds. Locations in BHF ponds where ^{22}Ne tracer was added or where samples were collected are indicated in Figure 3.2.

From a single station coinciding with the deepest point in the pond, tracer was injected into Pit S at a depth of 26 feet. The ^{22}Ne tracer was introduced as a gas, using a metal frit at the end of a tube, which produced small bubbles. Most of the Ne present in the gas would dissolve in the water during ascent of the bubbles. The amount of tracer that dissolved in the pond water was indicated by pond water samples (Table 4.10). The depth of tracer introduction was selected on the basis of dissolved oxygen, pH, conductivity, oxidation-reduction, turbidity, and temperature- all of which were measured every 3 to 9 feet of depth. The measurements of the aforementioned parameters indicated a possible thermocline-like zone of incomplete vertical mixing at a depth of 20 to 25 feet. Tracer samples were collected on the second, third, and fourth days of dosing in Pit S at two locations with two depths for each location. On September 2, tracer samples were collected at two depths and two locations in Lago los Osos, at one location (two depths) in Horseshoe Pond, at one location (two depths) in Rainbow Lake and at two locations (two depths) in the T pits. On the third day of dosing, Pit S was sampled at depths of 10 feet and 40 feet and analyzed for the full suite of water quality parameters.

Because all the pits in the Quarry Pits complex are interconnected by unrestricted, gravity-flow pipes or open channels, some of the ^{22}Ne added to (the Pit S portion of) Lago los Osos spread throughout the rest of the Quarry Pits complex. However, by preventing a significant through-put of water through the complex, the amount of tracer lost from Pit S to other ponds was minimized to the extent possible. But the inability to completely isolate surface water among the ponds in the Quarry Pits complex was the main reason why a single tracer was used for the BHF tracer experiment.

Groundwater monitoring at BHF wells commenced on August 24, 1998 and ended on December 8, 1999. Samples for tracer were collected generally on a weekly basis. Weekly water level measurements in various Newark Aquifer wells located throughout the BHF-portion of the study area were also recorded.

Alameda Creek Experiment

Xenon-129 was added to Alameda Creek on August 20 and 25, 1998. The segment of the creek that received tracer is in the AHF sub-basin and extends from the upstream side of Rubber Dam No. 3 to the Mission Boulevard Bridge (Figure 3.1). Tracer was added to the creek from a rubber raft at a position about 50 feet east of the rubber dam, near the middle of the channel, at a depth of about 5 feet. Because the maximum depth of water in the creek is shallow (limited by the height of the dam which is only six feet), vertical mixing was assumed during dosing and sampling. On both August 25 (the second day of dosing) and August 27, 1998, tracer samples were collected at two locations: one approximately five feet upstream of Rubber Dam No. 3 and the other upstream of the Mission Boulevard bridge (Figure 3.1). The samples were collected from a depth of 5 feet. On the final day of dosing, samples for the full suite of water quality parameters were collected on the upstream and downstream ends of the segment.

The plan at the time of dosing was to try to keep the ^{129}Xe tracer upstream of Rubber Dam No. 3 where there was less opportunity for surface water containing the tracer to migrate into recharge ponds through the diversion pipes. This strategy, if successfully carried out, would have allowed identification of the pathways in the aquifer followed by water recharging from this segment of Alameda Creek exclusively. However, some creek water containing ^{129}Xe was carried downstream of Rubber Dam No. 3 to water impounded by Rubber Dam No. 1 on the BHF side. This transfer of water, which occurred through a pipe that allows water to bypass Rubber Dam No. 3, was not desirable for the project but could not be avoided without an unacceptable disruption of ACWD's groundwater recharge operation. The amount of ^{129}Xe in surface (creek) water that may have entered the recharge ponds or resided in the pool of water immediately upstream of Rubber Dam 1 is not well known, but ^{129}Xe tracer was detected at a low concentration in Pit T on the BHF side.

Groundwater monitoring for creek tracer took place from August 1998 through March 1999. The analytical method used at Lawrence Livermore National Laboratory necessitated the collection of only one sample at a given monitoring point for analysis of each of the three tracers used in the AHF Sub-basin. Separate samples were not required.

Water Quality Characterization during Tracer Studies

Surface and groundwater chemistry was monitored concurrently with tracer, with the goal of determining possible changes in quality as a mass of water migrates from the pond to down-gradient wells. In so doing, it was hoped that gross non-flow related effects (transformation and/or retardation) could be discerned from flow-related (advection and dispersion) processes. Water quality parameters for this study included general minerals, alkalinity, pH, oxidation-reduction potential (ORP), conductivity, trace metals, total organic carbon (TOC), low-level volatile organic compounds (including MTBE), and coliphage (male-specific). The water quality portion of this study was conducted with a reconnaissance level of effort. Samples were generally collected on only one or two dates per pond or well with at least one sample collected at the time when the concentration of a tracer concentration in the well or pond reached its peak. Sampling was timed in this manner in order to ensure monitoring of the same 'unit' of water as it moved down-gradient. Inherent in this sampling strategy was the assumption that mixing and dilution with other sources of water (other than the ponds that were dosed with tracer) would be negligible. Pond samples were collected from two depths to represent deep and shallow pond water.

Aside from the isotopic tracers, no constituents, including coliphage, were artificially introduced into groundwater or surface water for this study.

Coliphage, a virus of coliform bacteria, is not harmful to humans, but can serve as an indicator of potential fecal contamination. Bacterial viruses are often used as non-pathogenic surrogates in experiments that require artificial introduction of a large concentration of viral particles into an aquifer. The underlying assumption in such experiments is that animal or human viruses would have similar fate and transport behavior under the same experimental conditions, and this assumption was implicit in this study.

SAMPLING EQUIPMENT AND PROCEDURES

Most pond samples were retrieved by use of a Kemmerer sampler (Wildlife Supply Company, Saginaw, MI. Model No. 1520-C20 0396). Mechanically operated from a boat, the Kemmerer sampler enabled collection of a discreet water sample at any desired depth in a pond. After being lowered to the target depth, the sampler was opened to allow water in, and then was sealed shut before being raised back onto the boat. Samples for low-level MTBE analysis and trace metals analysis were collected using a peristaltic pump. Tubing from the pump was acid-cleaned Teflon™ tubing, except for a short section of acid-leached silicon tubing at the pump heads.

Groundwater wells were purged and sampled with a portable two-inch, low flow-rate, stainless steel submersible pump. For each well, over three well volumes were purged prior to collection of samples. For visits involving only collection of samples for tracer or isotopic analysis, the pump was connected to ½-inch flexible PVC tubing. Teflon™-lined, ½ -inch tubing was used during visits to sample natural constituents (general minerals, trace metals, organics, etc), or combinations of natural constituents and tracer. To prevent cross-contamination in the field, the exterior of the down-hole equipment was sprayed with a solution of deionized water and tri-sodium phosphate in between individual wells.

Although no special field procedure was put in place to decontaminate the interior of the pump or PVC tubing prior to placement of the equipment down the well, the act of purging prior to sample collection should have achieved desired decontamination of the wetted interior of the equipment. There is no evidence that cross-contamination with respect to tracer constituents in the monitoring wells occurred. The progression of tracer detections from up-gradient to down-gradient wells followed a logical chronological sequence concluding with the detection in the production wells themselves. It should be noted that the production wells are equipped with dedicated pumps; therefore, their sampling did not involve use of tubing or the portable submersible pump. If cross-contamination among monitoring wells occurred, there would have been disproportional delays between detections in the production wells and the nearest up-gradient monitoring wells. Such delays were not seen. Moreover, cross contamination would have resulted in detections of ^{136}Xe (Kaiser Pond's tracer) in monitoring wells at Bunting Pond, or conversely, ^{124}Xenon (Bunting Pond's tracer) in monitoring wells next to Kaiser Pond. However, no such detections of multiple tracers were observed in any monitoring well.

Although no cross-contamination involving the use of PVC tubing to sample for tracers was observed, care was nevertheless taken during purging and sampling for other water quality parameters, which may have had the potential to adsorb to the inner walls of tubing. For field visits during which Teflon™-lined tubing was required, field staff were equipped with multiple lengths of pre-cleaned tubing, enabling fresh tubing to be used for each well on a given day of sampling. Upon returning from the field, the lengths of used Teflon™-lined tubing were decontaminated by immersion in a tub of tri-sodium phosphate solution.

Dissolved noble gas samples collected for isotopic analysis (targeting natural isotopes and tracers) were containerized in 3/8-inch copper tubes sealed with pinch clamps at both ends. Care was taken to purge the tubes, and to prevent the introduction of gas bubbles in the water samples intended for dissolved noble gas isotope analyses. This method of sample collection has a failure rate of less than 5 percent. In addition, a method was employed at well 21L03 to collect dissolved gas samples in situ, rather than at the surface using the pump. After purging at the desired depth, a nitrogen gas-filled copper tube with a check valve at the bottom was lowered to the sampling level. The tube was allowed to fill with water by adjusting the gas pressure. The tube was then pinched off with clamps at the surface. Well 21L03 was the only well where this method was employed because the well has three screened intervals to monitor shallow, middle, and deep parts of the aquifer to a maximum depth of 200 feet below ground surface (approximately 170 feet below the phreatic surface). This in-situ method was employed in the attempt to obtain samples representative of the formation at these discrete intervals of depth. The conventional method of purging and collecting samples of the pumped discharge would clearly have blended water from all three zones.

Pond and groundwater samples for trace metals were collected in acid-leached HDPE bottles, and were filtered and preserved with ultra pure nitric acid within 12 hours. Samples for oxygen and hydrogen isotopes were collected in 30-ml glass bottles, while samples for ^{14}Carbon were collected in 100-ml amber bottles with septa, and preserved with mercuric chloride. Samples for MTBE analyses were collected in pre-cleaned 40-ml VOA vials and analyzed within one week after sample collection.

Surface water samples collected from Kaiser and Bunting Ponds and all groundwater samples were field-measured for pH, temperature, dissolved oxygen, and conductivity using portable, hand-held electronic meters (Corning, Inc., Big Flats N.Y., Corning Code No. 475627; Orion, Inc., Beverley, MA, Model No. 250A). Oxidation-reduction potentials of groundwater samples were measured with a probe (Cole Palmer, Inc., Vernon Hills, IL Model 27013-41) that attached to the Orion meter.

August 1998 measurements in surface water for dissolved oxygen, oxidation-reduction potential, temperature, conductivity, and turbidity were performed in-situ with an electronic multi-parameter analyzer (YSI Inc., Yellow Springs, Ohio, YSI 6820 Multi-Parameter Water Quality Monitor). The analyzer relayed real-time data to the operator on the boat as it was lowered to the target depth, precluding the need to retrieve samples into the boat prior to measurement. However, samples for laboratory analysis had to be retrieved either with the peristaltic pump or the Kemmerer water sampler. Care was taken to calibrate the probes and hand-held meters in accordance with manufacturers' specifications.

Both the Cole Palmer probe and the YSI Multi-Parameter analyzer measure ORP by use of an Ag/AgCl reference electrode. The values reported in this study have been adjusted to be relative to 0 mV (i.e., a standard H_2/H^+ electrode.) by adding the reference potential of the Ag/AgCl electrode (approximately 190 mV).

All surface and groundwater samples collected for analyses of male-specific coliphage were sent to Biovir Laboratories in Benecia, Calif. Analyses of samples for xenon and neon tracers, other isotopic constituents, trace metals, and organics were performed at Lawrence Livermore National Laboratory. From the start of the project through July 1998, analyses of surface and groundwater samples for TOC and general inorganic water quality parameters (boron, general minerals, nitrate and nitrite) were performed by Sequoia Laboratories in Redwood City, Calif. Subsequent samples that targeted these parameters were analyzed by Clayton Laboratory of Pleasanton, Calif.

CHAPTER 4
RESULTS

NATURAL ISOTOPES, NOBLE GASES, AND GROUNDWATER AGE

Stable isotopes of oxygen, hydrogen, carbon (measured as the dissolved inorganic component, DIC), helium, and other noble gases, along with radiocarbon (measured as DIC) and tritium, were measured on water samples from production and monitoring wells.

Noble Gas Abundances and Recharge Temperature

Figure 4.1 includes plots of Xenon-132 abundance versus Argon abundance for various samples of ACWD surface water and groundwater. The temperature for a given water sample can be estimated by noting on the graph the position of the ^{132}Xe-Ar data point relative to reference lines corresponding to specific temperatures. Reference lines for 10 and 20 degrees C are included on Figure 4.1. An examination of ^{132}Xe and Ar abundances shows that in ACWD's production wells, recharge temperatures were recorded at approximately 14 to 15 degrees C, with little variation between wells. Surface waters plot in a tight group clustered around the temperatures measured in the field (10 to 11 degrees C for Bunting Pond and 12 degrees C for Kaiser Pond). Temperatures of groundwater samples are all higher than those of surface water samples. Values for well 21L08 are closest to surface water and show the most variation in temperature. Monitoring wells show more variability than production wells because they integrate a shorter water column and are more likely to retain a signal from short-term temperature variations. The approximate equivalence of recharge temperatures with local mean annual soil temperatures (http://www.giss.nasa.gov/data/update/gistemp/) indicates recharge is slow enough for temperature equilibrium to have been established between soil gas and recharging water (Stute and Schlosser 1993). The similarity of recharge temperatures (near the mean annual temperature) likely sets a scale for dispersive mixing in the aquifer, so that water several years old is really a mixture of ages spanning at least one year.

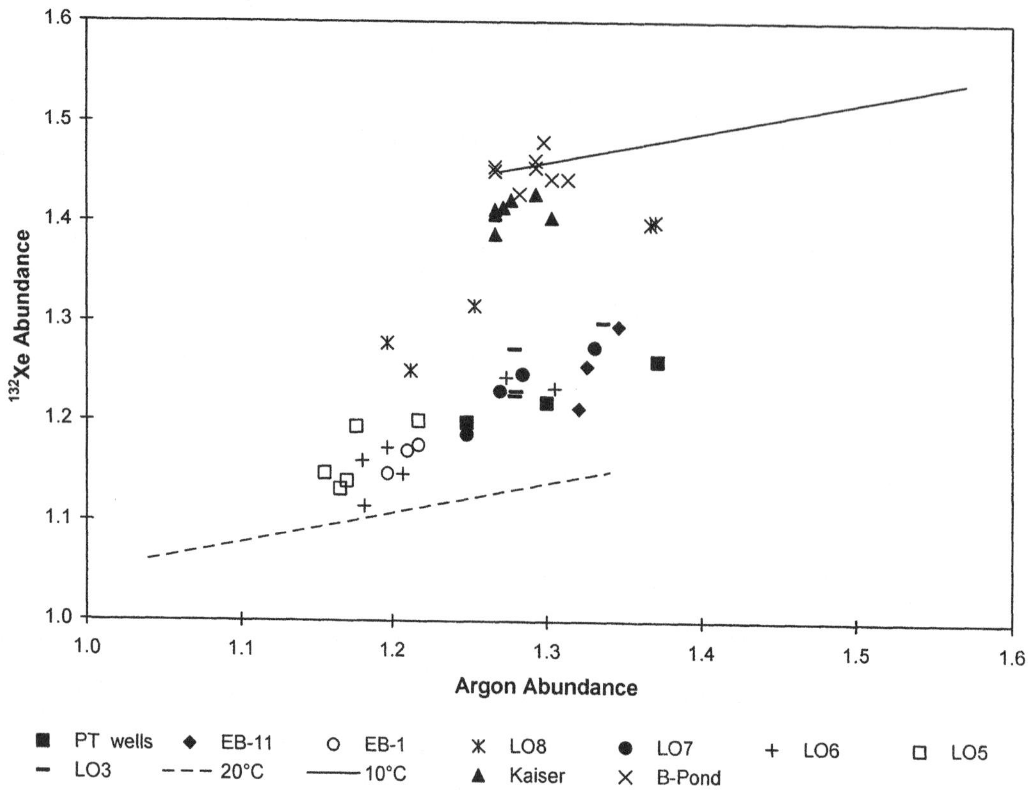

Figure 4.1 ^{132}Xe-Ar plot to determine water temperature in AHF Sub-basin at time of recharge. "Kaiser" and "B-Pond" refer to surface water in Kaiser Pond and Bunting Pond, respectively. EB-1, EB-11, L03, L06, L07 and L08 correspond, respectively, to wells 21J03, 21K05, 21L03, 21L06, 21L07, and 21L08.

An excess of atmospheric gases over equilibrium solubility concentration, is often found in natural waters, and is referred to as 'excess air'. Excess air is entrained in groundwater as the level of the water table fluctuates, and bubbles are partially dissolved in the recharging water. ACWD groundwater samples showed uniformly low values of excess air, an indication that variation between saturated and unsaturated transport is limited in extent and duration.

Ages

Ages were determined by the tritium-helium method described in Chapter 3. The amount of excess air in groundwater samples, a parameter that must be considered in the age calculations, was determined largely through the noble gas analysis as referenced in the preceding paragraph. The age-dating results are presented in Table 4.1 and in Figures 4.2, 4.3 and 4.4.

Groundwater ages were calculated for three production wells and two monitoring wells in the AHF sub-basin. These wells included PT-1, PT-4, PT-5, 21L03 and 27D08. While the mean ages of samples from all three depths in Well 21L03 were determined to be of less than one year, the down-gradient production wells have ages of 3.7, 4.7, and 8.8 years (analytical error ± 1 year). An age of 4.7 years was determined for Well 27D08, which, relative to the other four AHF wells, is more remote from the recharge facilities but has a shallower screened interval. If most of the inflow to the PT wells is derived from the recharge ponds, and if much of this flow occurs along a straight line instead of long circular route, then the average flow velocity from the recharge ponds to the production wells is approximately 300 ft/yr. This integrated flow velocity should not be confused with the minimum travel time. Furthermore, this does not take into consideration any water that is not recharged via the ponds. However, the mean of these ages, 5.7 years, is a rough indication of the residence time of the water in the sub-basin that is drawn into production wells. The range in ages (between <1 year and 8.8 years) indicates mixing between recently recharged water and ambient basin groundwater. The heterogeneity evident in the water ages is also reflected in the $\delta^{18}O$ and $\delta^{13}C$ values for AHF wells, which range from -5.1 to -6.7 $^{o}/_{oo}$ per mil and -11.7 to -15.7 $^{o}/_{oo}$, respectively (Table 4.1). These are larger ranges than one would expect to observe if, for example, there is a single recharge source and simple piston-type flow.

The ages in production wells in the Below Hayward Fault sub-basin are 3.0 years (Mowry 1), 5.5 years (Mowry 7), 19.7 years (Mowry 3), and 30.8 years (Mowry 2). This progression shows a pattern of increasing water age with depth of aquifer layer (Figure 4.4). Ages for several monitoring wells in the BHF sub-basin were estimated using the $^{3}He/^{4}He$ ratio from the samples and an assumed initial tritium value that is derived from the measured tritium values from other ACWD wells (the parameter $^{3}H_{o}$ in the equations of Chapter 3). To ensure validity of this "shortcut" method, ages calculated by the complete and "shortcut" methodologies (described in Chapter 3) for the Mowry production wells were compared and found to be in reasonably good agreement. Nevertheless, the quoted errors are greater for those wells whose water ages were determined by the shortcut method exclusively.

Table 4.1

Environmental isotope data

Sample location	Depth* (ft)	Date collected	$\delta\ ^{18}O$ (per mil)	$\delta\ ^{2}H$ (per mil)	$\delta^{13}C$ (per mil)	$^{3}He/^{4}He$	^{3}H (pCi/L)	Age (years)	^{14}C (pmC)	DIC (mg/L)
Above Hayward Fault										
21J03		12/16/97	-5.72	-47	-13.56	--	--	--	--	--
21J03		3/3/98	-5.95	-48	--	--	--	--	--	--
21K05		2/4/98	-5.54	-45	-12.83	--	--	--	--	280
21L03	80	12/2/97	-6.39	--	-14.40	1.005	--	1†	--	--
21L03	135	12/2/97	-6.31	--	-15.00	1.006	--	1†	--	--
21L03	135	3/3/98	-5.94	-47	--	--	--	--	--	--
21L03	185	12/2/97	-6.34	--	-14.30	1.008	--	1†	--	--
21L05		12/2/97	-5.41	-46	-14.13	--	--	--	--	300
21L05		2/4/98	-5.15	-46	-13.31	--	--	--	--	296
21L06		12/2/97	-5.34	-49	-12.80	--	--	--	--	302
21L06		2/4/98	-5.85	-48	-13.19	--	--	--	--	299
21L07		12/2/97	-5.76	-47	-11.73	--	--	--	--	292
21L07		2/4/98	-5.95	-48	-13.53	--	--	--	--	307
21L08		12/16/97	-5.64	-45	-12.35	--	--	--	--	260
27D08		1/6/98	-6.69	-53	-14.86	1.045	--	5†	--	663
Bunting West	20	12/30/97	-5.87	-48	--	--	--	--	--	--
Bunting West	5	12/30/97	-5.83	--	-11.22	--	--	--	--	245
Kaiser East	24	12/30/97	-5.71	-48	-11.91	--	--	--	--	252
PT-1		4/16/97	-6.15	-47	-15.68	1.146	48.2	8.8	88.4	481
PT-4		4/28/97	-5.73	-48	-14.47	1.037	31.9	4.7	92.6	384
PT-5		4/16/97	-5.92	-52	-14.80	1.018	25.3	3.7	95.4	330
PT6		6/9/98	-6.31	--	--	--	--	--	--	--
Below Hayward Fault										
20J06		8/24/98	--	--	--	1.077	--	5†	--	--
28D11		11/23/98	--	--	--	2.559	--	24†	--	--
28D12		11/24/98	--	--	--	1.065	--	5†	--	--
Mowry 1		6/30/97	-6.27	-48	-17.10	1.034	--	3†	97.8	354
Mowry 2		4/29/97	-7.21	-60	-13.39	3.427	94.0	30.8	84.2	293
Mowry 3		4/16/97	-6.47	--	-14.62	1.665	46.0	19.7	97.5	331
Mowry 7		4/16/97	-6.34	-50	-15.30	1.061	28.3	5.5	99.6	370
Mowry 9		12/16/98	--	--	--	1.648	--	19†	--	--
Mowry 9		6/8/99	--	--	--	1.706	--	20†	--	--

* Depth sample collected. Wells: below ground surface; ponds: below water surface.

† Ages are calculated using the tritium curve established with measured values on other wells. Errors are +/- 20% for one standard deviation.

Abbreviations: DIC= dissolved inorganic carbon; pmC= percent modern carbon.

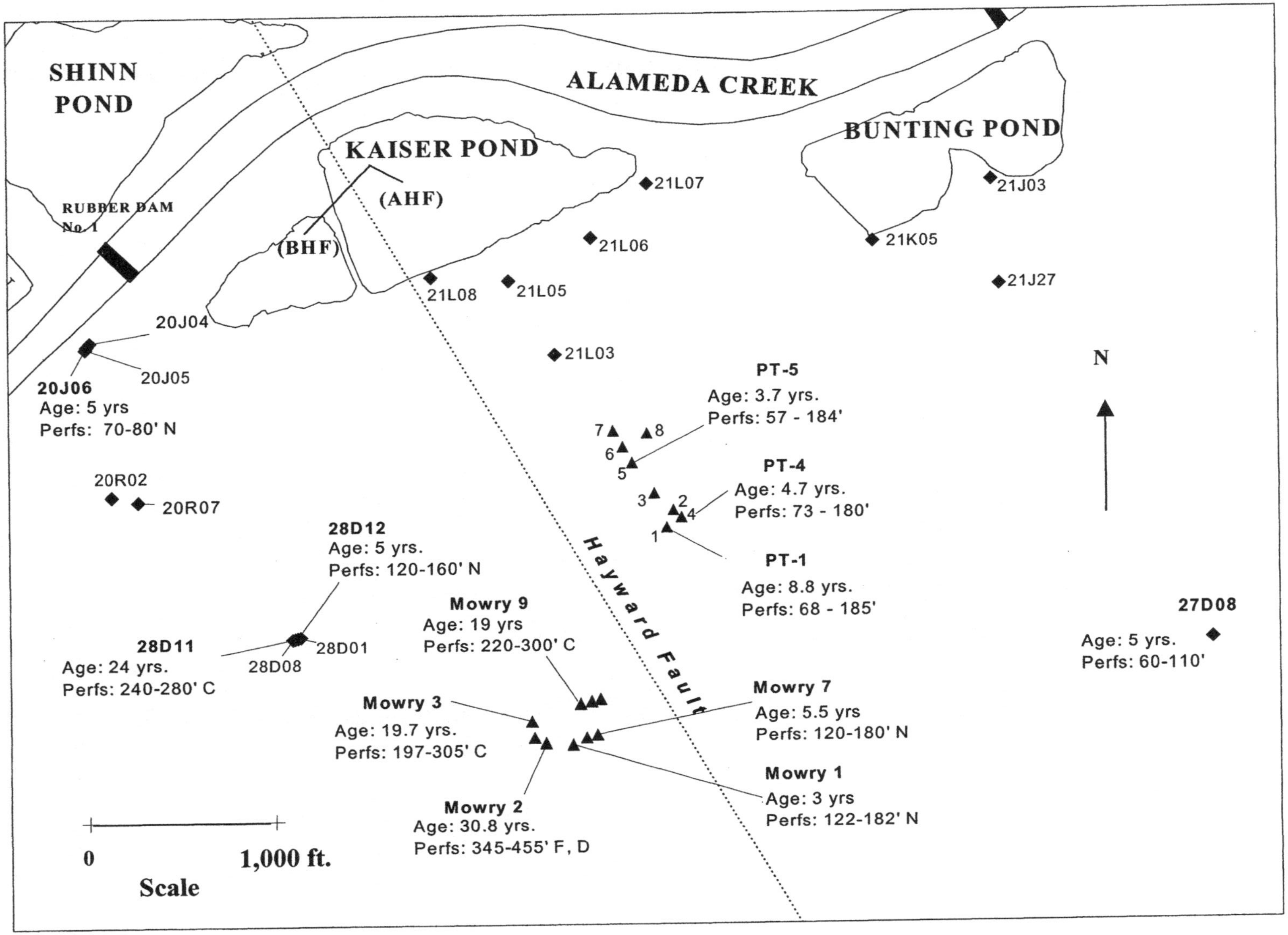

Figure 4.2 Groundwater ages by the helium-tritium method. The perforated interval of each well is indicated as feet below ground surface. BHF aquifer layers abbreviated as follows: N= Newark, C= Centerville, F= Fremont, D= Deep.

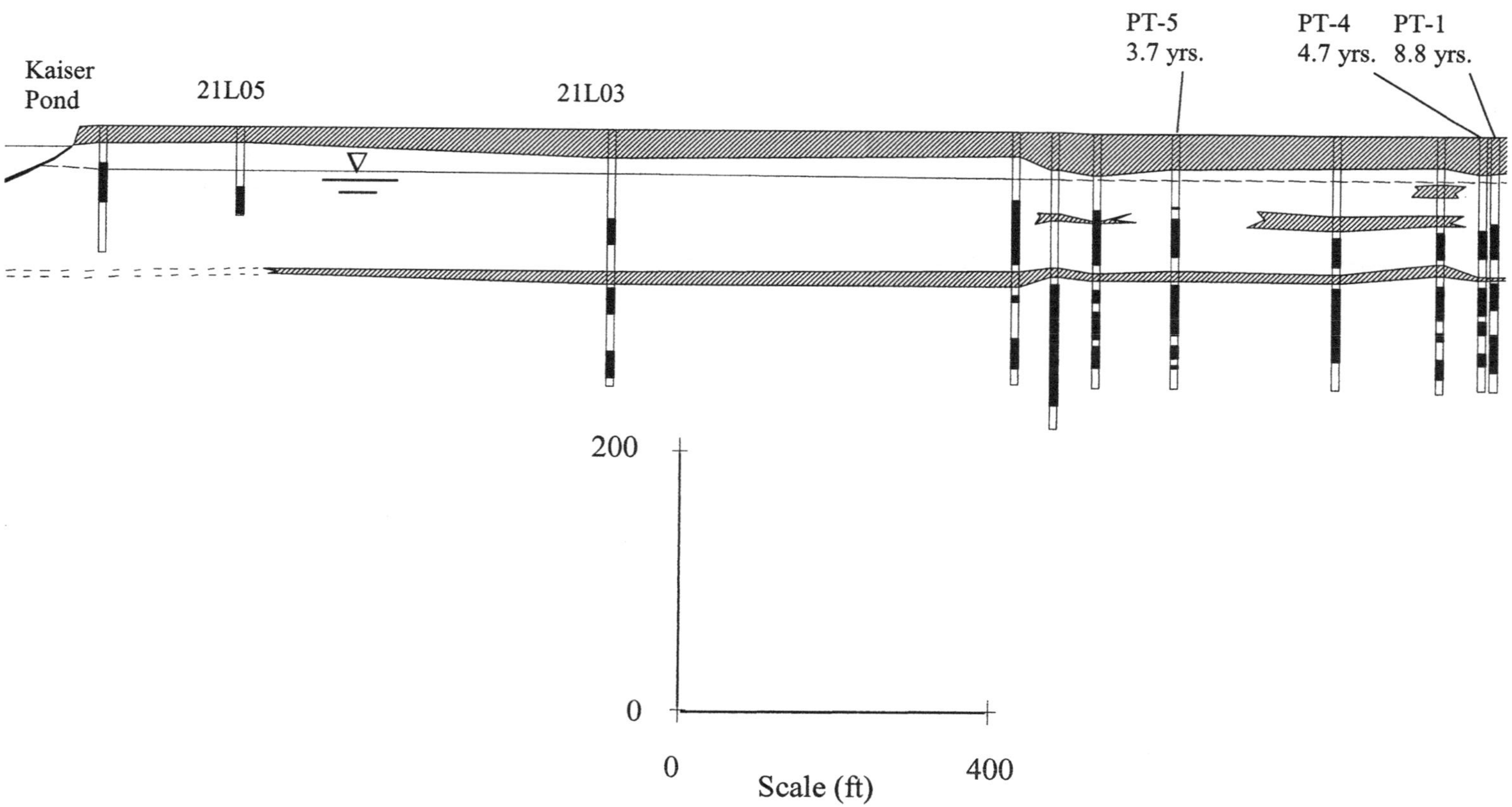

Figure 4.3 Age dating results for PT Wells (AHF), plotted in profile view along cross-section A-A' (refer to Figures 2.5 and 2.6). Darkened areas on the wells represent perforated/screened intervals.

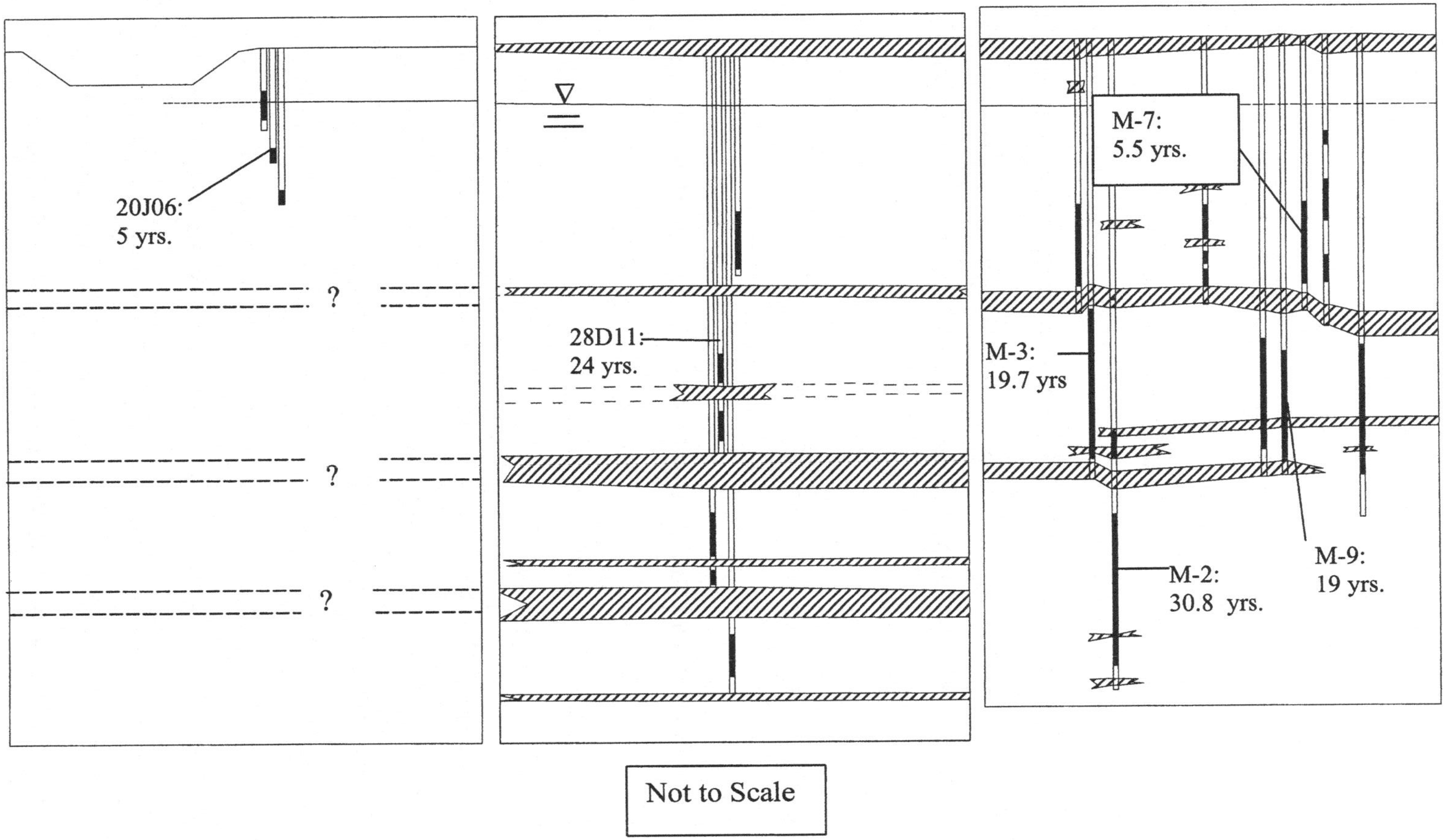

Figure 4.4 Age dating results for BHF wells, plotted in profile view. Each of three snapshots represents an individual segment of cross-section B-B' (refer to Figures 2.5 and 2.7). Darkened areas on wells denote perforated/screened intervals.

The range of groundwater ages for BHF production and monitoring wells span nearly the entire range that can be dated by the tritium-helium method. The oldest groundwater was found in Mowry 2. Its tritium level was 94 pCi/L, which corresponds to a decay-corrected tritium input level (that is, the initial tritium input, ^3H_o) of 450 pCi/L (not shown in Table 4.1). This value is consistent with bomb fallout-influenced precipitation for 1966. (Refer to Gat (1980) for a compilation of tritium level in precipitation over the period 1950-1970.) The relatively high tritium abundances in Mowry 3 and Mowry 2 suggest that the BHF sub-basin is less dispersive than the AHF sub-basin since dispersion would tend to reduce the tritium level by mixing with water of lower tritium abundance. The good agreement between the (decay corrected) measured tritium and the predicted tritium for the time of recharge (from the precipitation curve) indicate that mixing with pre-modern (pre-1950) water is not significant in these wells. The lack of variation in age values among multiple samples collected from the same aquifer also suggests a lower degree of dispersivity in the BHF sub-basin.

Deuterium and Oxygen-18

The $\delta^{18}O$ values show a correlation with groundwater age (where ages vary) and range from –6.3 $^o/_{oo}$ for Mowry 1 to –7.2 $^o/_{oo}$ for Mowry 2 (Table 4.1). This range is in the expected range for ambient groundwater recharged by water from the Alameda Creek watershed. While meteoric water that falls near the coast will have $\delta^{18}O$ values of –3 to –5 $^o/_{oo}$, meteoric water from Livermore, California (located 15 miles inland from this study area) has a mean, time-integrated $\delta^{18}O$ value of –7.7 $^o/_{oo}$. The $\delta^{18}O$ value for the recharge water will vary depending upon where in the watershed the bulk of the precipitation falls, the type of storm that brings the precipitation, and mixing processes before and during recharge. The trend toward more negative $\delta^{18}O$ values with increasing age may indicate increasing influence of imported water from a higher elevation source in the older, deeper waters (Figure 4.5 and Table 4.1). South Bay Aqueduct water has a $\delta^{18}O$ value of –9 to –10 $^o/_{oo}$ and is the only source of imported water from outside the watershed for the recharge facilities. The relative proportions of water from the two sources, the local Alameda Creek watershed and imports from the South Bay Aqueduct, vary substantially from year to year (Figure 4.6) and no clear trend with age emerges. If a significant fraction of water drawn from a well were South Bay Aqueduct water, the expected $\delta^{18}O$ value

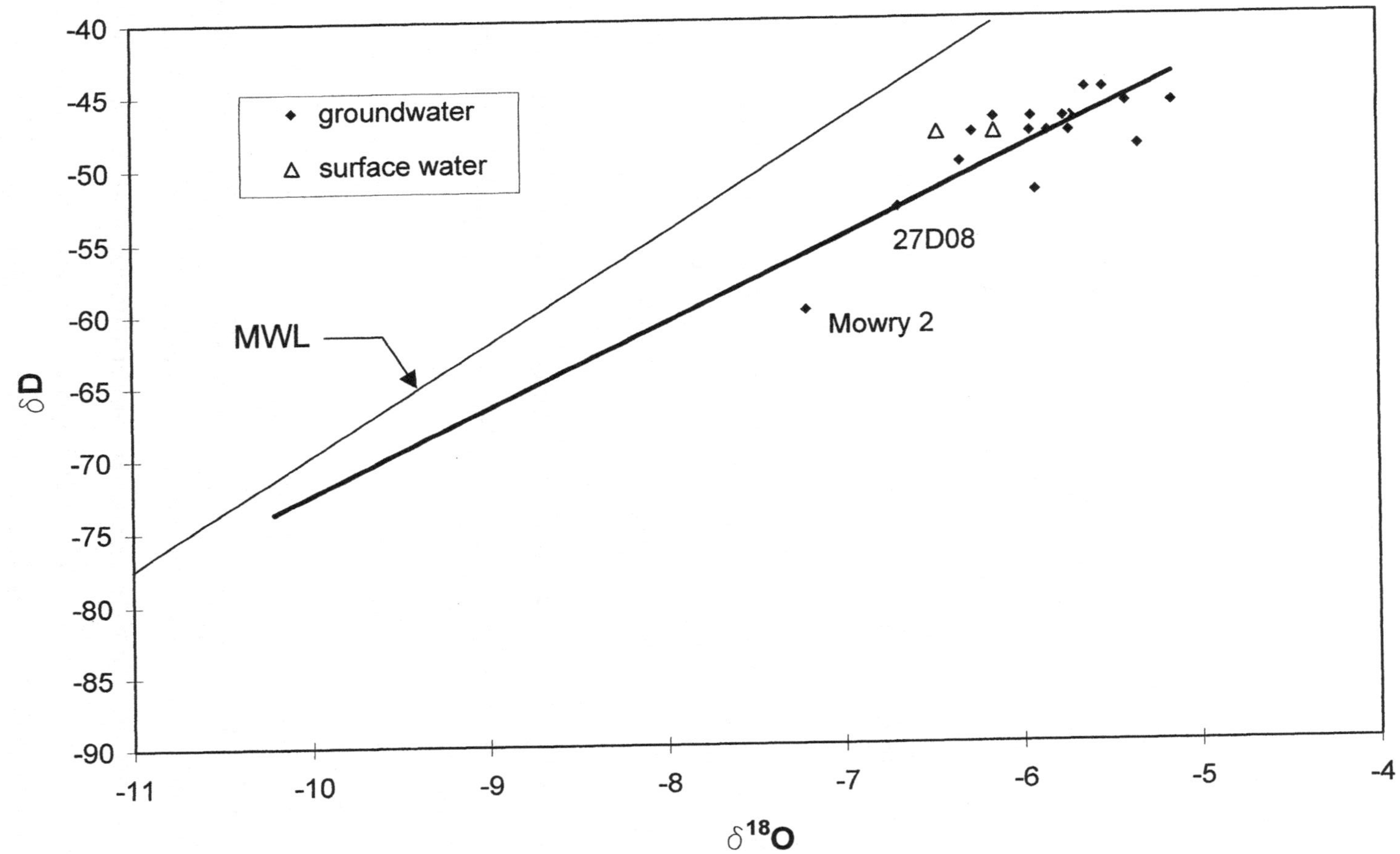

Figure 4.5 Stable isotope data with Meteoric Water Line (MWL)

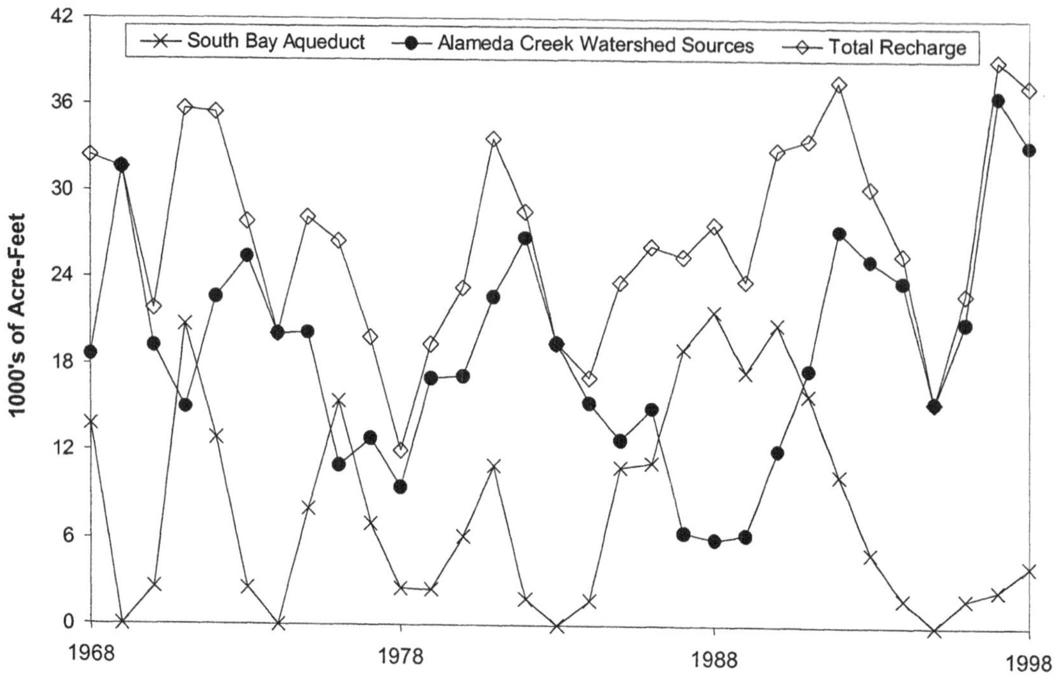

Figure 4.6 Sources of artificial recharge

would be less than –7.5 ‰. However, a more likely explanation for more negative $\delta^{18}O$ values in older water is that younger water has been evaporated to a greater degree than older water that was recharged before artificial recharge ponds were in operation. This is especially true for BHF water, as the large surface area of the ponds provides a greater opportunity for evaporation. Evaporation causes a shift in $\delta^{18}O$ to less negative values. The evaporative trend in groundwater is illustrated in Figure 4.6, whereby evaporated waters fall on a trend of lower slope than the Global Meteoric Water Line.

Carbon-13 and Carbon-14

Stable carbon (^{12}C and ^{13}C) and radioactive carbon (^{14}C) isotopes were measured on the dissolved inorganic carbon component in the water samples. Exchange of carbon with the various carbon reservoirs that the water comes in contact with can alter the $\delta^{13}C$ and the ^{14}C abundance of the carbon in the water. While biological processes such as photosynthesis and methanogenesis can alter the relative proportions of the carbon isotopes, by far the largest carbon reservoir with which the water exchanges is the carbonate minerals found in soils and aquifer material. In samples with significant variation, (given an analytical uncertainty of +/- 0.5 $^{o}/_{oo}$) the $\delta^{13}C$ values show a rough correlation with groundwater age as well, ranging from –17.1$^{o}/_{oo}$ in Mowry 1 to –13.4 $^{o}/_{oo}$ in Mowry 2 (Table 4.1). Dissolution of carbonate minerals (which have $\delta^{13}C$ values of approximately 0 per mil) along flowpaths can explain the increase in $\delta^{13}C$ values in water with an older component.

Levels of ^{14}C for the Mowry Wellfield range from 99.6 pmC (percent modern carbon) for Mowry 7 to 84.2 pmC for Mowry 2. This range is consistent with the interpretation that there is an older, more chemically evolved, groundwater component at depth (see Mook (1980) for a thorough description of sources of carbon in the subsurface and potential interactions between groundwater and aquifer materials). Present day surface water has ^{14}C levels of 100-110 pmC. Small amounts of bomb-derived or other technogenic ^{14}C can raise levels to above 100 pmC. Levels below 100 pmC are due to slow decay (half-life 5,730 years). In groundwater, interaction between dissolved inorganic carbon in water and soil and aquifer materials containing labile carbon forms can also bring ^{14}C below 100 pmC. Levels of less than 100 pmC in AHF wells corroborate that mixing with older water is taking place in production wells. The $\delta^{13}C$ values indicate that dissolution of carbonates is taking place on relatively short time scales.

In summary, ambient basin groundwater is dominated by an isotopic signature which has ^{14}C levels below 100 pmC, relatively negative values of $\delta^{18}O$ and $\delta^{13}C$, and relatively old ages. This water has interacted significantly with aquifer materials, and aside from high TDS, is of high quality (more detail on water quality is discussed later in this chapter). Recharge temperatures, determined from dissolved noble gas abundances, are uniform. This uniformity suggests that the time scale for dispersive mixing allows equilibration of recharging water with vadose zone gas. At the AHF production wells, recently recharged water with a distinct isotopic signature and young age is mixed with this ambient water due to movement through a highly dispersive groundwater environment. Source related heterogeneity in water from the AHF sub-basin is indicated by variation in groundwater ages and stable isotopes from the Peralta-Tyson Wellfield. In contrast, there is significantly less mixing between ambient groundwater and recharge water in the BHF aquifers, particularly in the deeper Centerville-Fremont and Deep Aquifers.

HYDROLOGIC CONDITIONS DURING THE STUDY PERIOD

Surface Water Elevations

Hydrographs of pit elevations for ponds that are closest to the wellfields are provided on Figure 4.7. As indicated in the figure, the water level in AHF ponds varied by less than several feet during the study period, with Kaiser Pond fluctuating between 45 and 52 feet, and Bunting Pond between 59 and 64 feet MSL. BHF Ponds, Pits S, T1, and T2 demonstrated much greater fluctuation, particularly Pit S with a range of 14 feet MSL to 35 feet MSL. In the early part of the study, prior to April 1998, surface water levels below the fault were low (approximately 15 feet MSL) following an earthwork project in Lago los Osos (which includes Pit S), Horseshoe Pond, and Rainbow Lake. By May 1998, however, these ponds were re-filled to near capacity (capacity in these BHF ponds is 35 feet MSL) as a consequence of a very wet winter generating substantial runoff in the Alameda Creek watershed for diversion into the ponds. BHF pit levels dipped in the late summer of 1998, but rebounded in the winter of 1999, closely following a

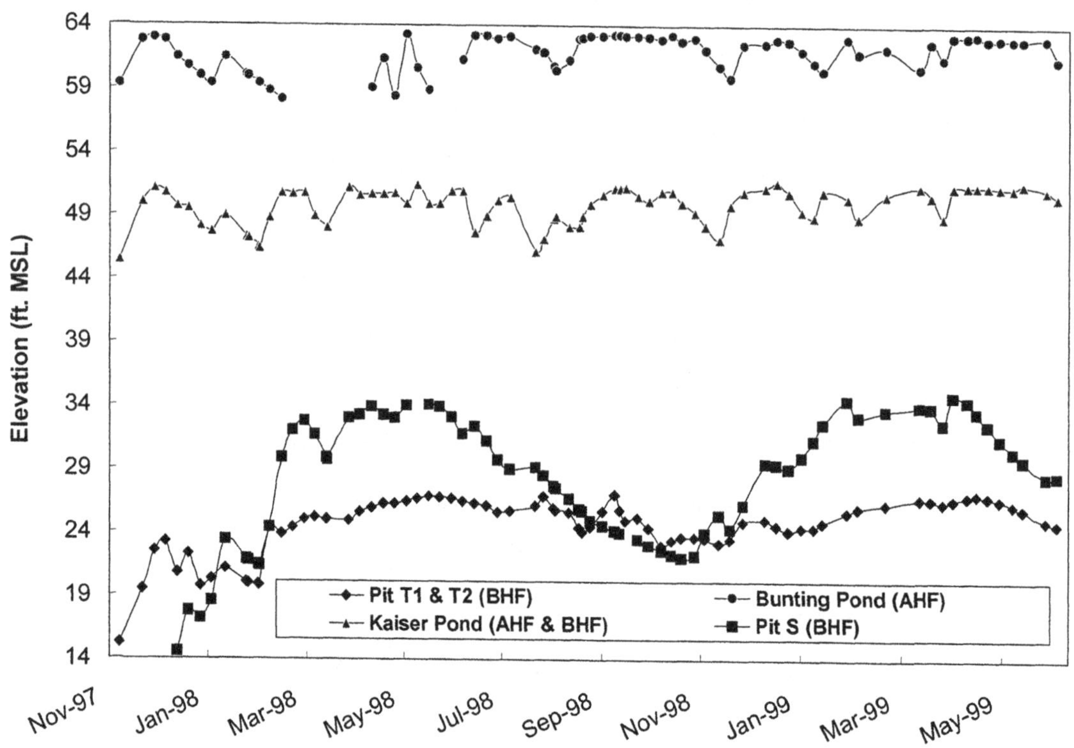

Figure 4.7 Hydrographs for selected recharge ponds

typical seasonal pattern of wet winters with significant creek flow and low groundwater pumping and dry summers with reciprocal conditions. The typical seasonal pattern of fluctuation in BHF ponds following the spring of 1998 was unrelated to the earthwork project, which had been completed in the summer of 1997.

The water level fluctuations in AHF ponds did not follow this seasonal pattern because of continued inputs of surface water in the summer months. ACWD's operating policy gives priority to AHF ponds in the allocation of the small amount of dry-season creek flow. BHF ponds receive water in the summer only after AHF recharge facilities are full (during wet months, there is typically plenty of creek flow to fill recharge ponds on both sides of the fault). Hence, BHF pond levels tend to dip in the summer while AHF ponds stay nearly full year-round. The AHF sub-basin has fairly low groundwater storage per unit change in peizometric head. Therefore, ACWD must continuously feed the AHF recharge ponds in order to maintain acceptable well levels at the Peralta-Tyson Wellfield. Maintaining dynamic pumping levels above the top of the perforations helps to avoid corrosion and prevents entrainment of air in the discharge water. The BHF sub-basin has a lot more storage per unit change in peizometric head, and it is usually not difficult to maintain acceptable well levels at the Mowry Wellfield in dry months.

The pit levels are provided in tabular format in Table A-1 in the appendix.

Wellfield Pumping

Monthly pumping from individual Mowry and PT production wells are provided in Tables 4.2a and 4.2b, respectively. Totals for the two wellfields, as provided in the last column of each table, indicate that summer and winter were high and low demand seasons, respectively, for each wellfield. The peak-demand period for the wellfields was June to October. Well demands are directly dictated by demands on ACWD's distribution system, which in turn, are driven by weather and temperature. Demands rise sharply during hot, dry weather. In the late fall, winter, and early spring within the study period, PT Wellfield demand varied from approximately 300 to 800 acre-feet per month. During the hotter, drier months, production typically ranged between 1,000 and 2,000 acre-feet per month. At the Mowry Wellfield,

Table 4.2a

Monthly totals of PT Wellfield (in AHF Sub-basin) pumping during study period (acre-feet)

Month	PT-1	PT-2	PT-3	PT-4	PT-5	PT-6	PT-7	PT-8	Total
Sep-97	193	287	366	0	208	334	351	0	1738
Oct-97	316	369	280	0	0	172	129	0	1264
Nov-97	2	65	131	0	0	28	109	0	335
Dec-97	37	53	120	0	0	72	17	0	299
Jan-98	182	0	27	0	0	193	146	16	564
Feb-98	158	0	347	0	0	0	307	0	811
Mar-98	144	4	220	0	0	0	163	0	531
Apr-98	37	44	294	0	0	0	225	0	599
May-98	124	73	277	0	0	32	0	0	506
Jun-98	233	124	281	0	59	128	163	0	986
Jul-98	101	0	203	190	213	285	314	134	1439
Aug-98	124	0	227	281	276	273	222	285	1689
Sep-98	30	0	152	113	169	252	279	73	1067
Oct-98	41	0	261	193	280	278	269	0	1323
Nov-98	154	0	197	278	64	47	79	0	818
Dec-98	2	0	88	124	5	81	10	0	311
Jan-99	1	0	72	9	8	197	55	0	341
Feb-99	5	212	11	22	59	66	85	0	460
Mar-99	42	117	28	55	76	13	43	104	477
Apr-99	16	64	32	193	180	77	79	181	821
May-99	182	232	258	272	222	239	248	286	1939
Jun-99	94	167	257	232	259	212	271	237	1729
Jul-99	251	204	258	242	251	253	281	265	2005
Aug-99	170	159	261	242	254	170	151	194	1600
Sep-99	101	66	164	124	304	188	55	211	1213
Oct-99	248	177	190	194	351	231	40	271	1702
Nov-99	131	206	206	186	164	0	0	214	1107
Dec-99	72	83	157	134	160	46	0	120	772

production was typically less than 500 acre-feet in the off-peak season and fluctuated between 500 to 1,000 acre-feet per month during the peak-demand season.

Groundwater Piezometric Head

Groundwater levels in various wells were recorded to help in the interpretation of the direction of groundwater flow throughout the study area, and to assist in understanding surface water/groundwater interactions. Although most monitoring wells were targeted for well level measurements, water levels were recorded in only one production well from each of the two

Table 4.2b

Monthly totals of Mowry Wellfield (in BHF Sub-basin) pumping during study period (acre-feet)

Month	Mowry 1 (Newark)	Mowry 2 (Fremont, Deep)	Mowry 3 (Cent.)	Mowry 6 (Cent.)	Mowry 7 (Newark)	Mowry 9 (Cent.)	Total
Sep-97	84	0	317	43	177	157	778
Oct-97	91	0	327	103	203	253	977
Nov-97	51	0	329	1	98	10	488
Dec-97	81	0	156	1	96	124	458
Jan-98	33	0	73	1	138	2	246
Feb-98	39	0	276	14	199	133	660
Mar-98	22	0	125	0	106	12	265
Apr-98	14	0	144	2	64	15	238
May-98	59	0	118	1	244	91	513
Jun-98	51	0	307	4	243	124	729
Jul-98	69	0	335	2	277	72	754
Aug-98	92	0	321	1	246	39	699
Sep-98	84	0	268	54	280	134	820
Oct-98	87	0	248	44	280	117	776
Nov-98	67	0	3	1	208	222	501
Dec-98	46	0	4	1	90	111	253
Jan-99	68	0	0	1	68	248	383
Feb-99	83	0	0	0	228	77	388
Mar-99	52	0	0	1	195	131	380
Apr-99	62	0	0	1	195	125	38
May-99	5	0	0	141	9	303	457
Jun-99	61	0	0	164	219	353	796
Jul-99	97	0	0	221	154	384	855
Aug-99	99	0	0	172	0	382	654
Sep-99	35	37	204	54	3	361	695
Oct-99	96	0	104	16	1	190	407
Nov-99	63	0	268	21	253	139	743
Dec-99	53	0	198	3	66	49	369

Note: Mowry wells 4, 5, and 8 were not active during the study period for this project.
Abbreviations: Cent= Centerville (Aquifer)

wellfields. The two production wells were PT-8 (AHF) and Mowry 8 (BHF). Unlike most of the other production wells, these two wells were inactive for most of the study period, enabling static water level data to be collected from them. Moreover, Mowry 8 is also screened in the Newark Aquifer, the only BHF aquifer expected to contain detectable levels of tracer within the timeframe of the study.

A complete data set of groundwater piezometric head is provided in Table A-1 in the appendix. Groundwater elevations recorded on September 22, 1998 and December 8, 1998 are illustrated in plan view in Figures 4.8 and 4.9. The profile of 12/8/98 water levels along cross-sections A-A' (AHF) and B-B' (BHF) is provided in Figures 2.6 and 2.7, respectively. Surface water elevations are also included for comparison. Each figure gives a snapshot-in-time corresponding to times of moderately high (9/22/98) and low groundwater (12/8/98) production. PT-8 was not running during June through early September of 1998 and therefore was able to serve as a monitoring point for static head within the wellfield. Mowry 8 was not pumping on either 9/22/98 or 12/8/98.

The sharp contrast in water level in AHF ponds relative to nearby wells, as seen in Figures 2.6, 4.8 and 4.9, indicates that the groundwater phreatic surface may not rise to coincide with the pond water surface at the pond perimeters in the AHF Sub-basin. The highly transmissive AHF aquifer appears to "disperse" pond seepage toward the wellfield at a rate appreciably faster than the rate of seepage from the pond. Hence, groundwater gradients between the wellfield in the ponds are relatively flat and pond water is, to a certain extent, perched above the phreatic surface. Below the fault, however, the groundwater phreatic surface may rise to meet the pond elevation surface. Groundwater and surface water level data suggest that groundwater did not rise to coincide with creek surface water on either side of the fault on 9/22/98 or 12/8/98. Moreover, the elevations of groundwater in nearby wells never rose higher than the bottom of the creek at their lowest points. As indicated in Table 2.2, the elevation of the bottom of the creek at Rubber Dams 1, 2 and 3 are 43, 30, and 53 feet MSL, respectively. Groundwater levels at nearby wells (all of which are along the south bank) did not rise higher than these elevations; however, well 20J04 did rise as high as 30 feet MSL in April 1999. Hence, the creek near Rubber Dam 2 may have become a gaining stream in the spring of 1999. Upwelling at lower, downstream segments may have started earlier in the year, potentially inducing a shift in the direction of groundwater flow away from the Mowry Wellfield during a period of time within the study period.

A comparison of clustered wells with different screened intervals (20J04, 20J05, 20J06 and 20R03, 20R04, and 20R05), demonstrates that a strong vertical component to groundwater flow within the Newark aquifer exists near the BHF pits. For a given cluster, piezometric head correlates to the elevation of the screened interval of the wells. Shallower-screened wells had higher piezometric heads and deeper-screened wells had lower piezometric heads. For example, the top-of-screen depths of 20J04, 20J05, and 20J06 are 30, 99, and 70 feet below ground

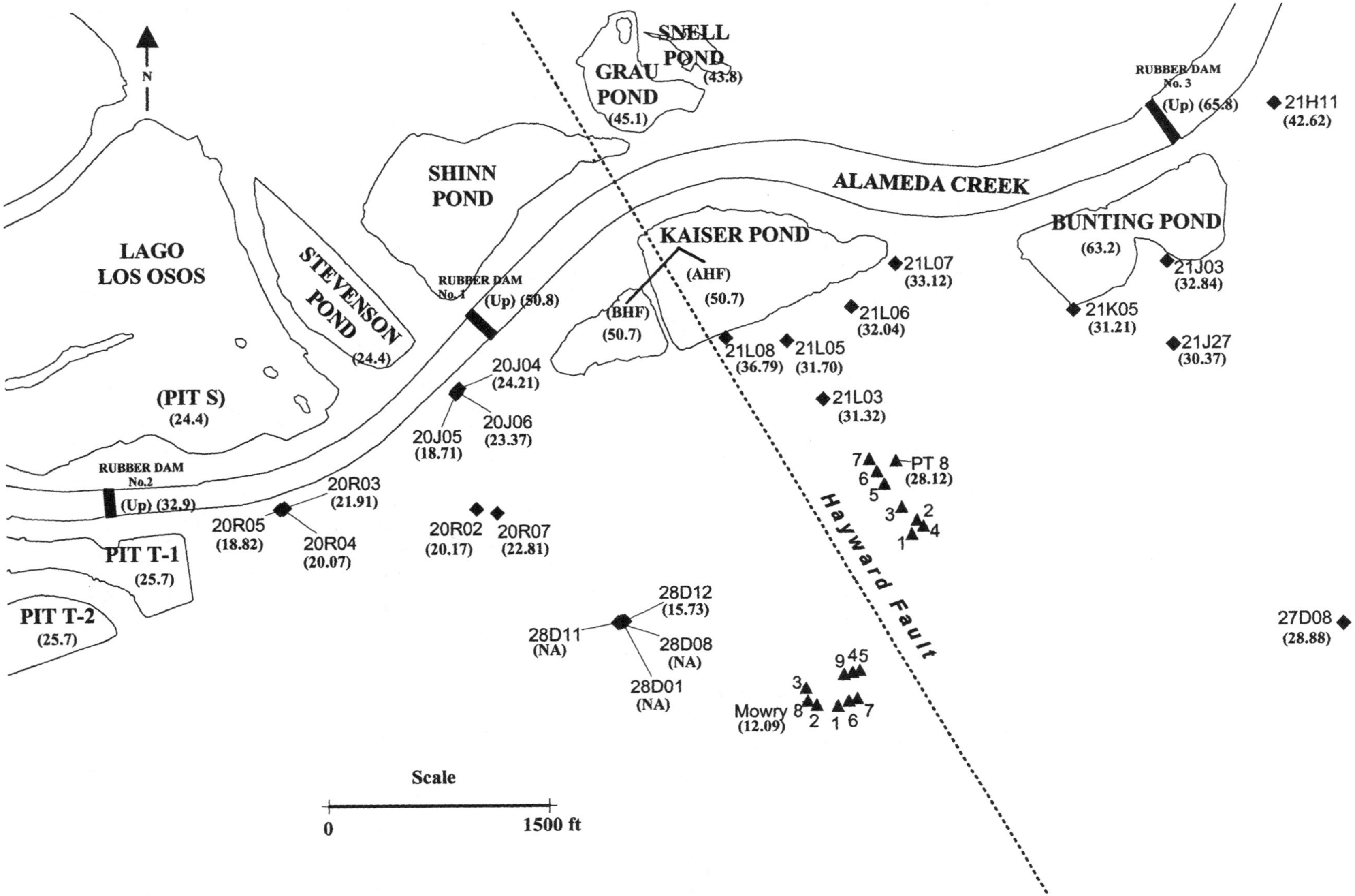

Figure 4.8 Elevations of surface water and groundwater piezometric head on 9/22/98 (ft, MSL)

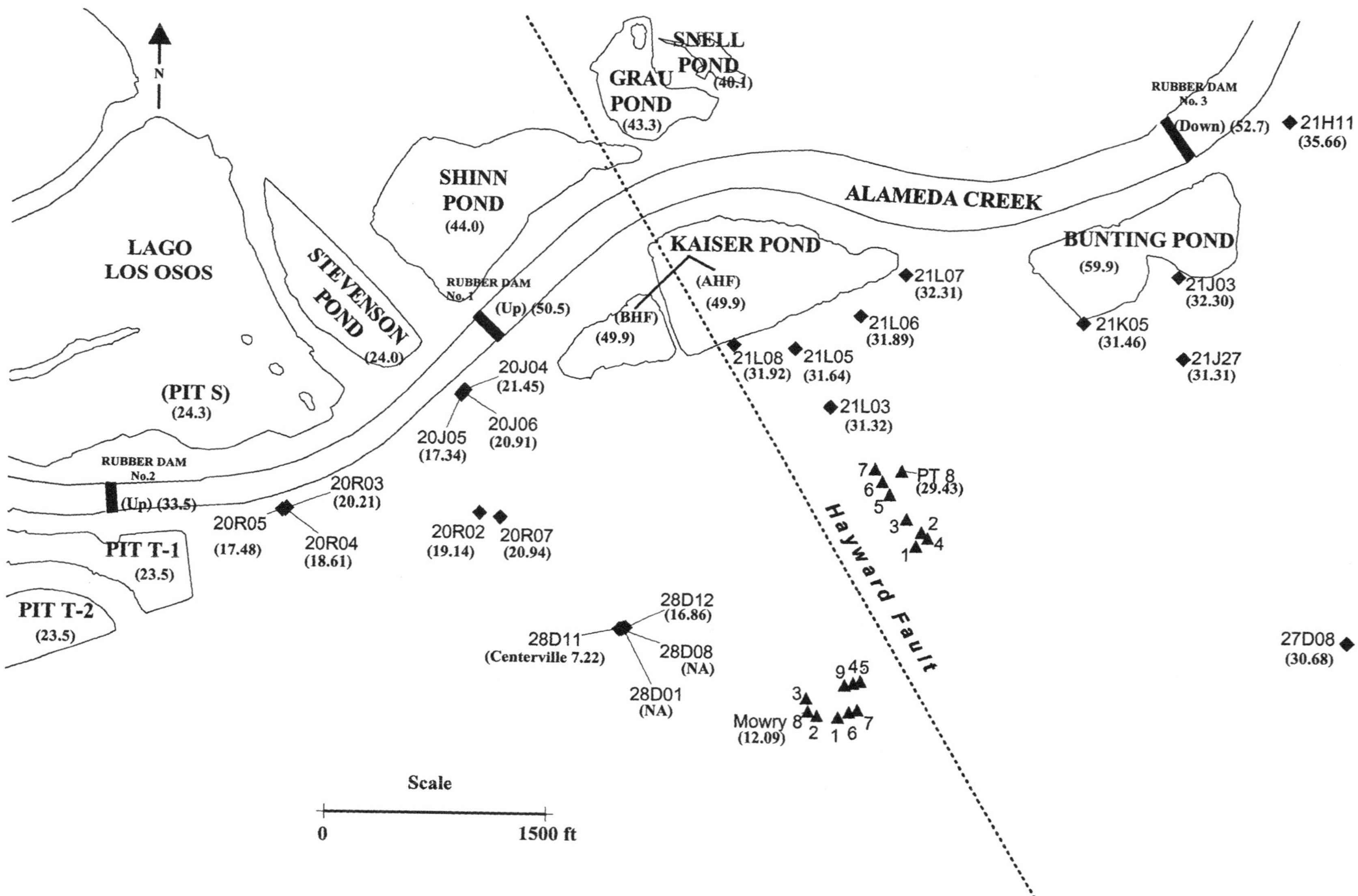

Figure 4.9 Elevations of surface water and groundwater piezometric head on 12/8/98 (ft., MSL)

surface, respectively (Table 2.3). As indicated on Figure 4.9, piezometric heads for these three wells were 21.45, 17.34, and 20.91 feet MSL, respectively, on 12/8/98. Figure 4.10 shows that this difference among the wells within the cluster was generally maintained throughout the study period.

This observed downward vertical-flow component is likely a consequence of the wells' close proximity to the recharge facilities and the non-uniformity of seepage rate with respect to depth in the aquifer. Because of the shallow depth of Alameda Creek and the probability of a clogging layer in the deep areas of the adjacent recharge ponds, most of the recharge water is probably introduced into the Newark Aquifer at elevations near or above the phreatic surface. Introduction of recharge water primarily at shallow depths could give rise to a downward vertical flow component. Other factors that could drive downward flow include aquifer heterogeneity and leakage to deeper aquifers.

Characterization of piezometric head in this part of the study area (immediate proximity to the recharge facilities) is therefore a three-dimensional problem, and additional clusters of wells with short screens would have been beneficial. If the vertical flow component attenuates with distance away from the ponds, it may be valid to use water levels in the available wells (shallow-screened wells near the creek and deeper-screened wells away from the creek) to

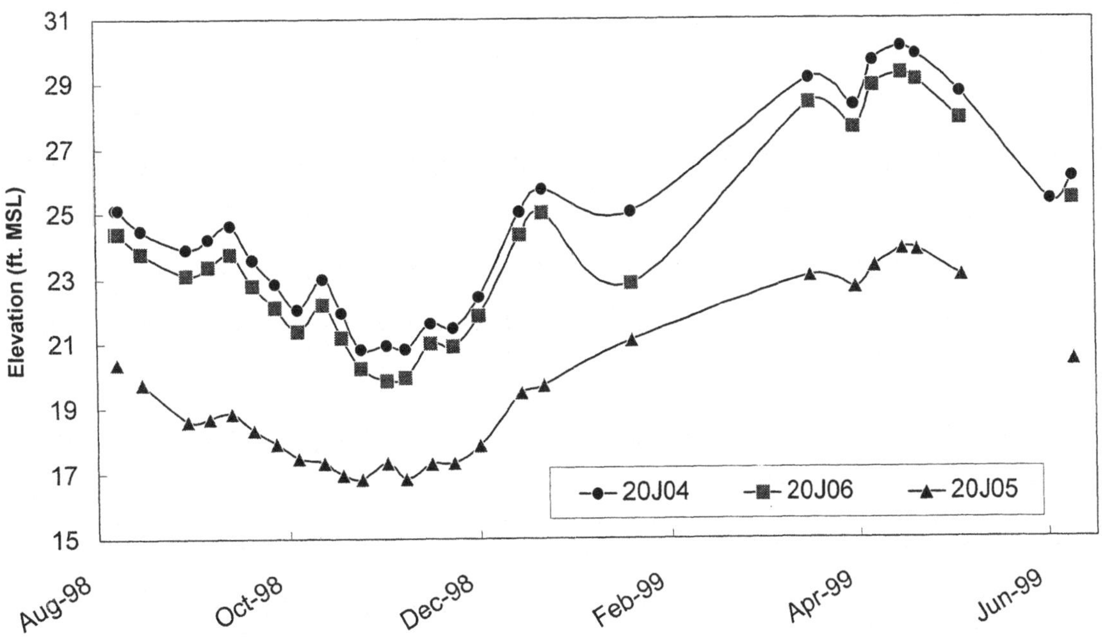

Figure 4.10 Piezometric head in selected BHF cluster wells screened in Newark Aquifer (ft. MSL)

calculate gradients for streamlines along the phreatic surface. A comparison of water level hydrographs for wells 20J04, 20R07, 28D12 confirmed the expected time-consistent pattern of decreasing heads from BHF recharge facilities to the Mowry Wellfield (Figure 4.11 (a)). Hydrographs of AHF wells 21L05, 21L03 and PT-8 also show the same time-consistent pattern (Figure 4.11 (b)). The reader should be aware that PT-8 was pumping on certain occasions within the time period represented by this hydrograph, and therefore, was not representative of static head at the PT Wellfield on such occasions. It should also be noted that PT-8 is screened strictly below the thin fine-grained soil layer that runs through the PT Wellfield, and hence, is not necessarily representative of the piezometric head above the layer.

Despite the limitations of the well network, the analysis was taken one step further with calculations of horizontal gradients for the AHF Aquifer and the Newark Aquifer (the upper aquifer on the BHF side), using data appearing on Figures 4.8 and 4.9. Table 4.3 compares the gradients and pumping data recorded on 9/22/98 and 12/8/98. Mowry 8 and PT-8, both considered in the calculation of gradients, were not pumping on these two dates.

The gradients in Table 4.3 were calculated in linear fashion by simply dividing the water level differential by the horizontal distance between the two wells in each pair. Admittedly, these calculated gradient values are not ideal in a Darcy calculation of flow velocity for the following reasons. Firstly, the overall well network was not designed to characterize flow in a three-dimensional flow environment. Secondly, flow near the wellfields is quasi-radial, and this should impart non-linearity to the peizometric surface. Thirdly, according to geologic data (see Figures 2.6 and 2.7), the Newark Aquifer (the upper aquifer on the BHF side) is unconfined in the study area, as is the upper 115 feet of the AHF aquifer. This lack of confinement should also impart some curvature to the piezometric surface as recorded in the wells. Finally, on the BHF side, the alignment of 20J04, 20R07, 28D12 and Mowry 8 appears to coincide with a cross-gradient vector- not the actual down-gradient vector.

Despite these shortcomings, the gradient calculation is useful in contrasting the impact of wellfield pumping on groundwater heads within the study area. The results provided on Table 4.3 indicate a correlation between gradient and pumping in the AHF sub-basin, with higher pumping inducing higher gradients and lower pumping inducing lower gradients. Also, higher gradients were generally observed in the BHF sub-basin during the higher demand period. However, the water level differential between 28D12 and Mowry 8 was actually greater in the

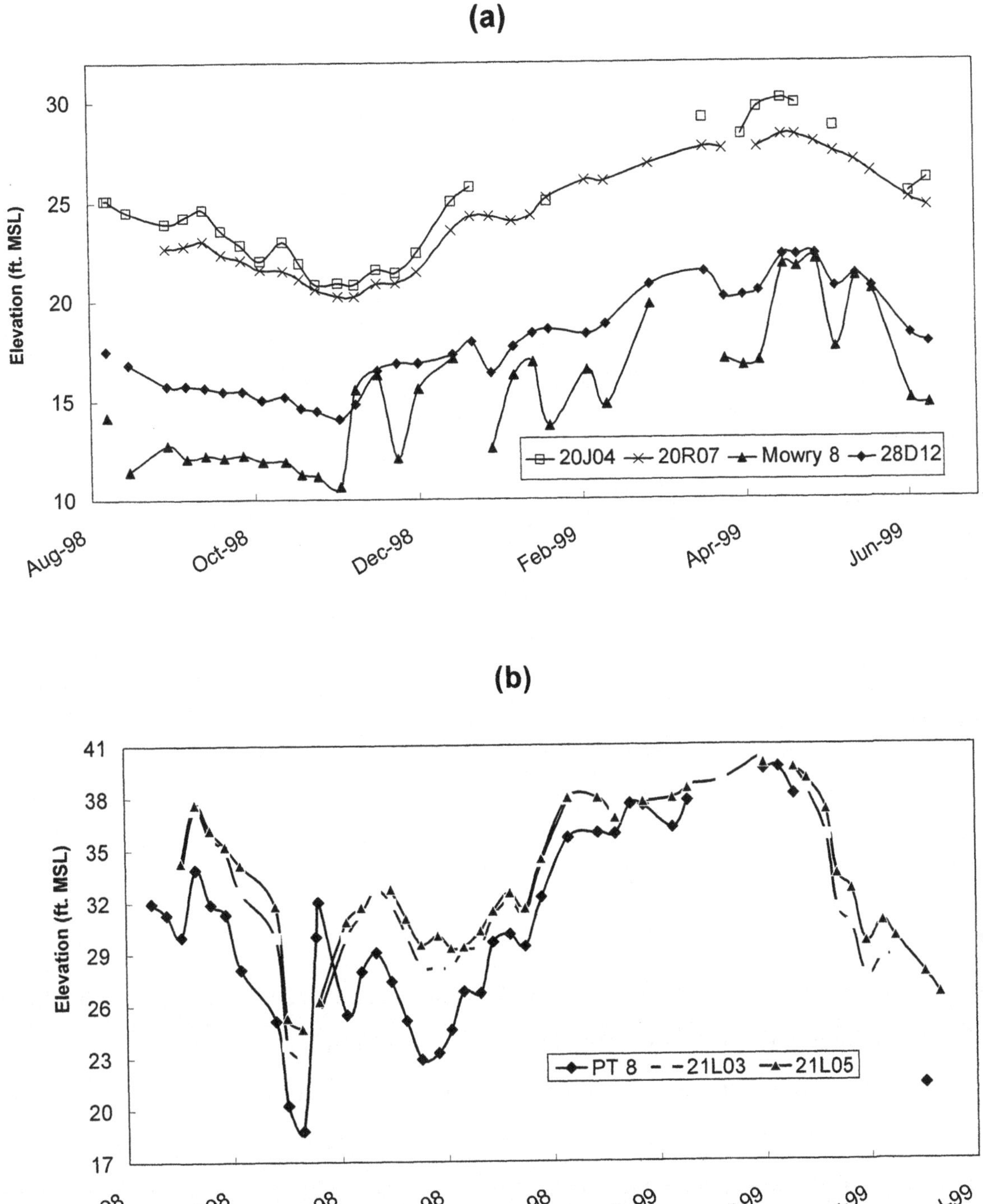

Figure 4.11 Hydrographs for selected wells (a) BHF (b) AHF

Table 4.3

Comparison of gradients during periods of high and low rate of wellfield production

		9/22/98		**12/8/98**	
Above Hayward Fault					
PT Wellfield production rates		(ac-ft/day)		(ac-ft/day)	
1 day average		39.8		20.2	
avg. prev. 15 days		38.3		13.0	
avg. prev. 30 days		39.9		20.0	
Selected wells for gradient calculation	Horizontal distance (ft)	Head difference (ft)	Gradient (ft/ft)	Head difference (ft)	Gradient (ft/ft)
21L05 to 21L03	475	0.38	0.0008	0.32	0.0007
21L05 to PT 8	1118	3.58	0.0032	2.21	0.0020
21K05 to PT 8	1601	3.09	0.0019	2.03	0.0013
Below Hayward Fault					
Mowry Wellfield production rates (Newark Aquifer wells only)		(ac-ft/day)		(ac-ft/day)	
1 day average		11.8		0.0	
avg. prev. 15 days		11.8		0.8	
avg. prev. 30 days		11.6		6.2	
Selected wells for gradient calculation	Horizontal distance (ft)	Head difference (ft)	Gradient (ft/ft)	Head difference (ft)	Gradient (ft/ft)
20J04 to 20R07	908	1.40	0.0015	0.51	0.0006
20R07 to 28D12	1140	7.08	0.0062	4.08	0.0036
28D12 to Mowry 8	1373	3.64	0.0027	4.77	0.0035

lower demand period (12/8/98) than in the higher demand period (9/22/98), suggesting that head in 28D12 is not strongly influenced by pumping at the Mowry Wellfield.

In contrast to hydrologic conditions below the Hayward Fault, the influence of wellfield pumping on groundwater near the recharge facilities appears to be strong. Piezometric head as recorded in monitoring wells, including those closest to the recharge ponds, rises and falls inversely with pumping rates (Figure 4.12).

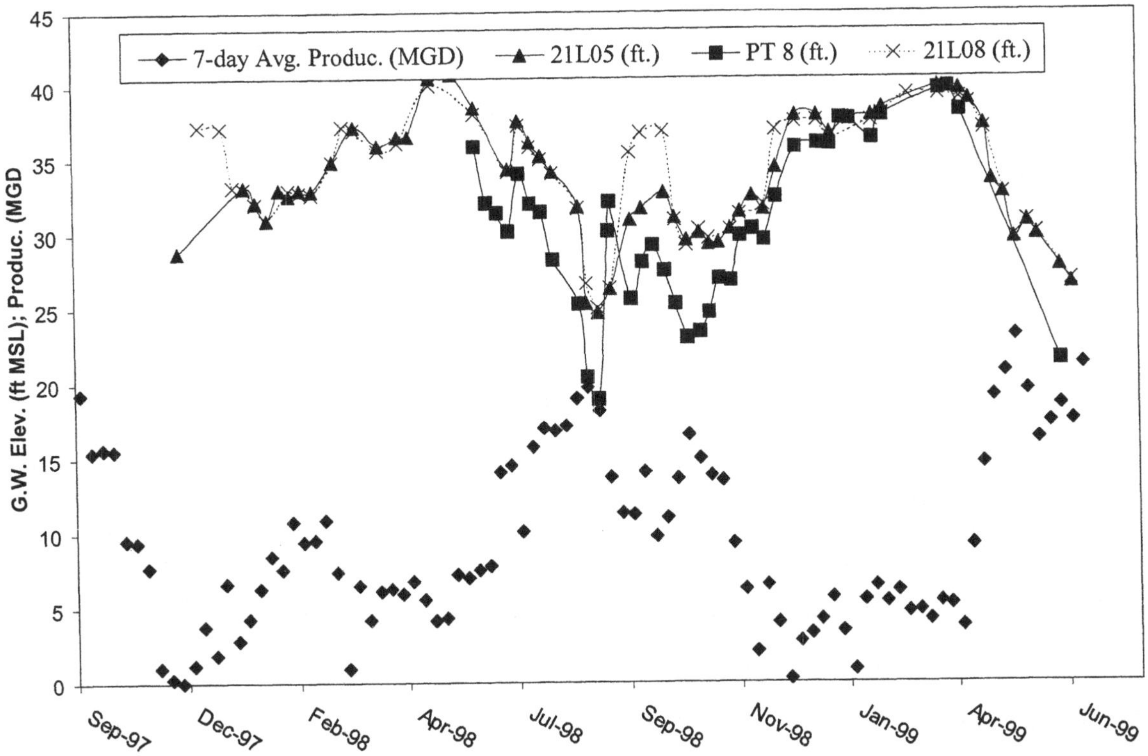

Figure 4.12 Comparison of PT Wellfield pumping and groundwater piezometric head in AHF Sub-basin

Investigation of Seepage from Kaiser Pond (AHF and BHF)

There has been substantial research on hydraulic phenomena that influence the rate at which surface water seeps from lakes, ponds and infiltration basins. Bouwer (1994) is recommended as a source for the reader interested in an in-depth discussion on this topic. Further research in this area was not an objective of this study. However, in order to interpret and explain the tracer experimental results, the authors opted to investigate, at least on a reconnaissance level, site-specific factors that controlled the migration of tracer from the ponds. Kaiser Pond was targeted for a seepage evaluation because of the availability of daily measurements of pond levels and surface water inflow needed to calculate seepage rates (see Chapter 3 for the methodology). Seepage rates were compared with pond levels, groundwater heads in nearby AHF and BHF monitoring wells, and differences between pond and groundwater levels (Figure 4.13). Groundwater head on both sides of the Hayward Fault (AHF well 21L08 and BHF well 20J04) was considered because Kaiser Pond spans the Hayward Fault and pond water probably seeps to both the AHF and BHF sub-basins. The levels in Kaiser Pond-BHF are the same in Kaiser Pond-AHF except when the level drops below the culvert that interconnects the two ponds. Pond levels remained above the culvert for the duration of the study period; therefore, for purposes of this analysis, the two ponds were at the same level.

As indicated in Figures 4.13, the correlation between seepage and the above-mentioned variables varied considerably, with r-squared values ranging from nearly zero to 0.7. Among the parameters investigated, seepage rates correlated most strongly with pond inflow rates (Fig. 4.13 (a)). One caveat to this apparent strong correlation is that seepage rates were not determined independently from inflow rates. There is some concern, therefore, that a high correlation coefficient for these two parameters could have been predestined. Any errors in the inflow rate would have produced similar-valued errors in the seepage rate, and such close tracking would have acted to give a high correlation coefficient. Despite this limitation in the analysis, the lack of correlation between seepage rates and pond stage, and between seepage rates and aquifer head, is noteworthy. It is also interesting that the correlation was poor for aquifer head on both sides of the Hayward Fault. Correlation coefficients for seepage versus head-related variables may have been greater had the data set focused on periods when inflow to the pond was zero. As indicated in Figure 13(a), seepage rates were not zero during time steps in which inflow rates were zero.

The results of this analysis, in conjunction with the observed discontinuity between surface water and groundwater levels (see Figure 2.8 and compare Kaiser Pond with well 21L08), suggest that Kaiser Pond seepage is controlled by a clogging layer covering the bottom and submerged banks of the pond. According to Bouwer (1994), the development of clogging layers in infiltration basins are "the rule rather than the exception", and may consist of inorganic or organic materials. Sedimentation, precipitation, and bacterial secretions may drive the formation of this layer. Bouwer also discusses the importance of draining infiltration basins and enabling clogged surfaces to dry as a means of restoring infiltration capacity.

The pattern of behavior indicated in Figure 4.13, with seepage well correlated to inflow but poorly correlated to head-related variables, suggests time-dependent variability in clogging layer permeability. This variation, if it exists, could be a consequence of intermittent wetting and drying within the range of pond operating levels. The authors hypothesize that surface water diverted into Kaiser Pond percolates fastest through the portion of the side-slope that was above

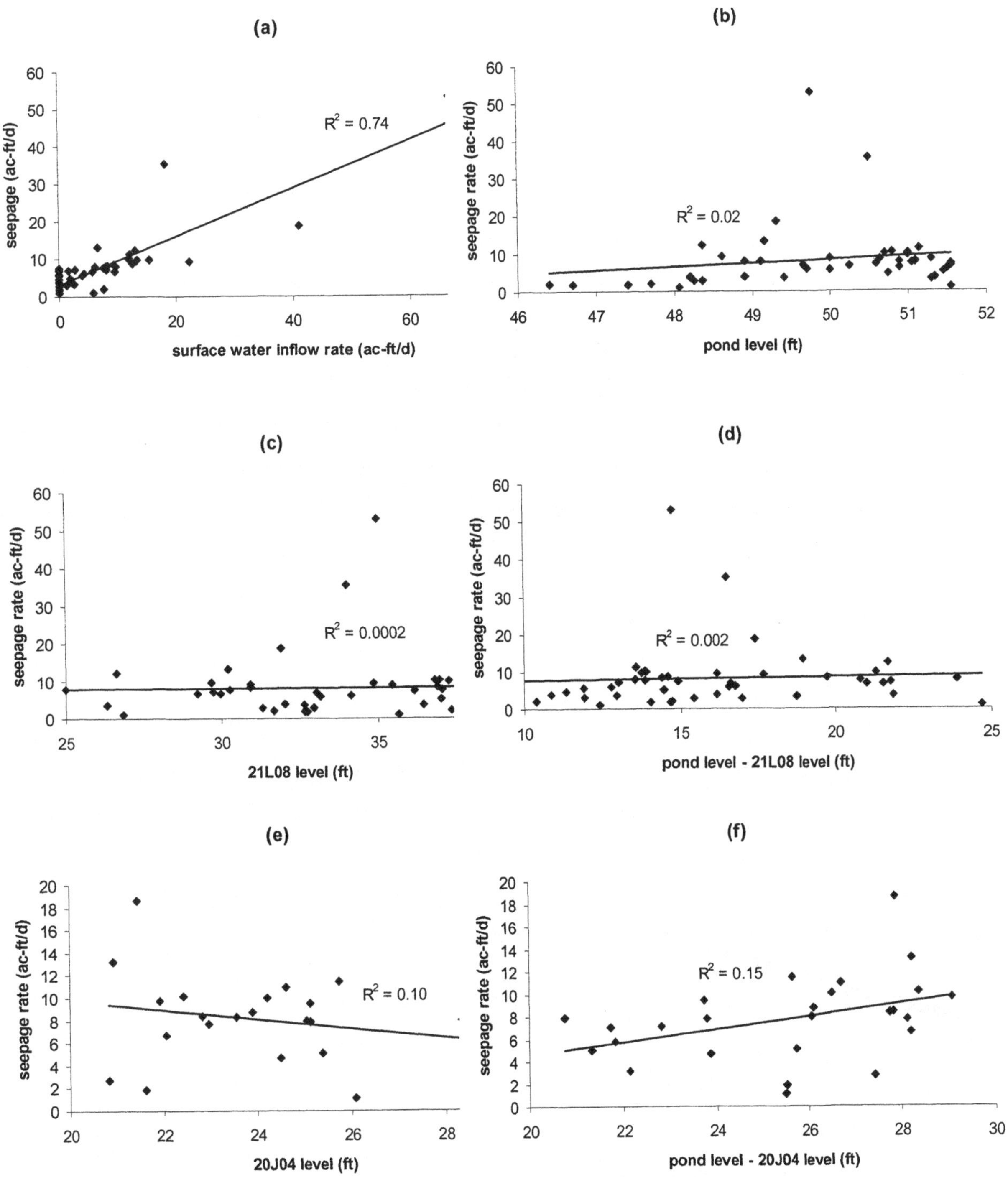

Figure 4.13 Kaiser Pond seepage vs. (a) inflow to pond, (b) pond water level, (c) water level in well 21L08 (AHF), (d) head difference between pond and well 21L08, (e) level in well 20J04 (BHF), and (f) head difference between pond and well 20J04. Water levels in ponds and wells are expressed in feet mean sea level.

the pond level (and exposed to air and sunlight) just prior to the inflow event. Over time, re-submergence may re-establish clogging and reduction in seepage rates. But after the water level drops and the previously submerged part of the bank is exposed to air and sunlight, the clogging materials may break up, restoring permeability to some degree. The length of drying time prior to re-submergence may be an important contributing factor toward restoration of permeability in the slopes of Kaiser Pond. In the deep, constantly submerged parts of the pond, clogging layers are probably fairly thick and well established. Seepage rates at such depths are probably much lower. Hence, recharge to the aquifer probably occurs mostly at very shallow depths in the pond, as indicated graphically in Figure 4.14. The permeability of near-shore areas that are not constantly submerged, and therefore subject to periodic drying, may be sufficiently high to enable percolation to generally keep up with inflow from Alameda Creek. But the clogging layer in the constantly submerged parts of the pond appears is evidently pervasive enough to enable pond levels to be maintained at elevations significantly higher than surrounding groundwater,

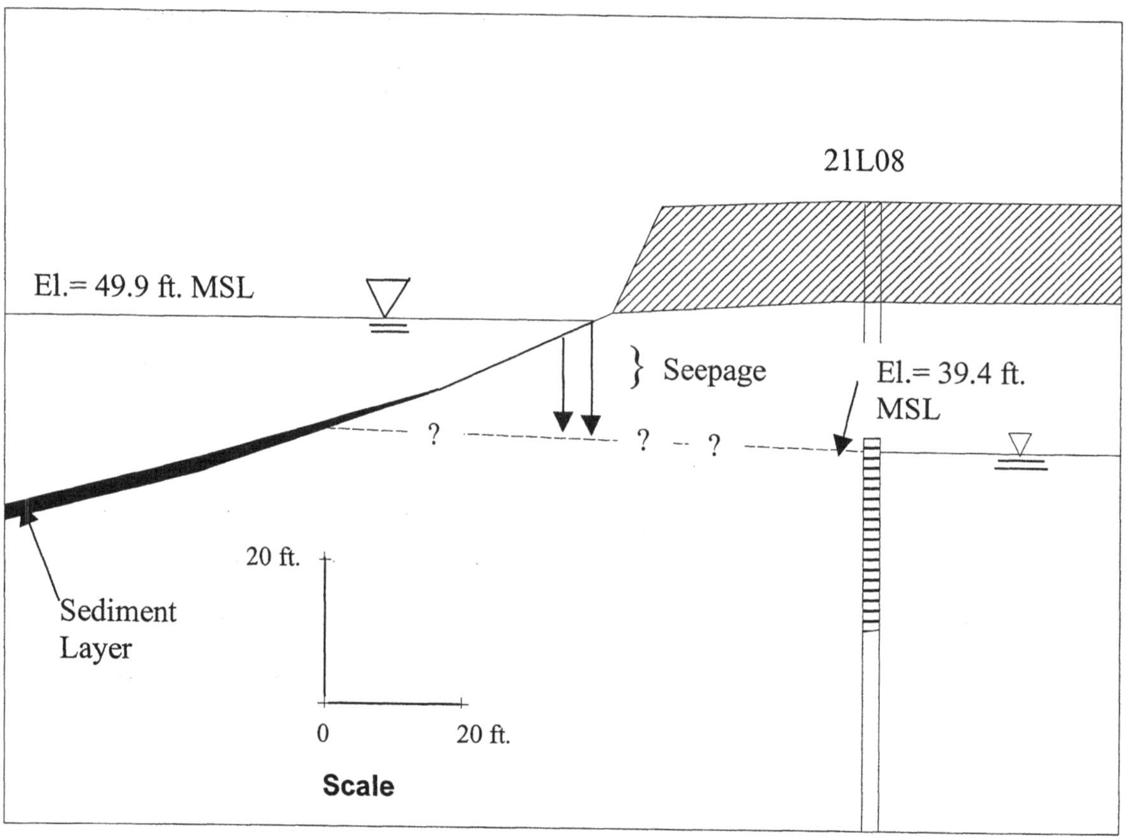

Figure 4.14 Profile of Kaiser Pond near well 21L08 (AHF). Includes elevation of surface and groundwater as measured on 12/8/98. Thickness of clogging layer, as depicted here, is for illustrative purposes only and not based on actual measurements. Diameter of well 21L08 diameter is exaggerated. Cross-hatched area represents fine-grain soil.

and to prevent hydraulic influences in the aquifer, such as wellfield pumping, from strongly affecting pond seepage. However, as indicated in Figure 4.12, once a water particle leaves the pond and arrives in the AHF aquifer, it is then subsequently subject to the strong influence of the Peralta-Tyson Wellfield.

Bunting Pond could have been a better choice for this seepage analysis because it is sufficiently distant from the Hayward Fault and, like Kaiser Pond, is not subject to surface water transfers with other ponds. However, the meter on the inflow pipe to Bunting Pond was not functioning during much of the study period, and sufficient data could not be generated. The findings for Kaiser Pond should be applicable to Bunting Pond since the water level in Bunting Pond is similarly perched above the groundwater phreatic surface.

The mechanics of seepage and the behavior of the aquifer system have important implications on the interpretation of tracer migration after dosing AHF ponds. For an average tracer molecule to reach the aquifer, starting from a large depth in the water column near the center of the pond, dispersion or diffusion would have had to carry it upward and laterally toward very shallow near-shore areas. From there, it probably seeped to the aquifer in response to the influx of surface water into the pond from Alameda Creek. Once in the aquifer, subsequent migration toward the wellfield was most likely driven by pumping at the Peralta-Tyson Wellfield.

Direct characterization of the clogging layer, including measurements of thickness and permeability, was outside the scope of this study.

AHF POND TRACER STUDIES (^{136}Xe AND ^{124}Xe)

Results for the AHF tracer study are tabulated in Tables 4.4 and 4.5. Twenty-two surface-water samples from Kaiser and Bunting Ponds, and 137 groundwater samples from AHF monitoring and production wells, were analyzed for Xe isotopes. The concentration of tracer in each sample is expressed as a normalized concentration C/C_o, which is the concentration of tracer in the sample divided by the estimated peak concentration in shallow groundwater in immediate proximity to the pond. For ^{136}Xe, the maximum concentration measured in well 21L08 was chosen as C_o, since this value is within a few percent of the values measured in the shallow region of Kaiser Pond on 12/29/97, and comes from water known to have recharged to

Table 4.4

^{136}Xe tracer in Kaiser Pond, AHF monitoring wells, and AHF and BHF production wells

Sample location/ description*		Sample date	$^{124}Xe/^{132}Xe$	$^{136}Xe/^{132}Xe$	^{136}Xe (cc (STP)/ g H_2O)	^{136}Xe C/C_0 (%)
Kaiser-East at 10'	SW	12/29/1997	0.03469	0.34546	5.37E-11	80.31
Kaiser-East at 15'	SW	1/13/1998	0.00431	0.36599	1.25E-10	186.94
Kaiser-East at 30'	SW	1/6/1998	0.00362	0.36460	1.17E-10	174.97
Kaiser-East at 35'	SW	12/29/1997	0.03143	0.34640	5.70E-11	85.24
Kaiser-East at 35'	SW	1/26/1998	0.00362	0.36654	1.26E-10	188.43
Kaiser-Middle at 4'	SW	1/26/1998	0.00358	0.34522	5.19E-11	77.62
Kaiser-Middle at15'	SW	12/23/1997	0.00380	0.33888	3.08E-11	46.06
Kaiser-Middle at 15'	SW	1/13/1998	0.00433	0.36489	1.20E-10	179.46
Kaiser-Middle at 30'	SW	1/13/1998	0.00449	0.36602	1.25E-10	186.94
Kaiser-Middle at 33'	SW	1/26/1998	0.00356	0.36643	1.25E-10	186.94
Kaiser-West at 7'	SW	1/6/1998	0.00361	0.35972	1.02E-10	152.54
21L03	MW	12/29/1997	0.00361	0.33004	-2.82E-15	0.00
21L03	MW	1/27/1998	0.00354	0.33011	2.03E-13	0.30
21L03	MW	2/10/1998	0.00357	0.32999	-1.60E-13	-0.24
21L03	MW	2/10/1998	0.00359	0.33009	1.55E-13	0.23
21L03	MW	2/17/1998	0.00350	0.32972	-5.25E-13	-0.78
21L03	MW	2/17/1998	0.00360	0.32992	-5.89E-13	-0.88
21L03	MW	3/10/1998	0.00358	0.32909	-2.85E-12	-4.26
21L03	MW	3/31/1998	0.00358	0.32912	2.00E-13	0.30
21L03	MW	6/30/1998	0.00355	0.33022	1.31E-13	0.20
21L03	MW	1/12/1999	0.00353	0.33034	2.00E-13	0.30
21L05	MW	12/2/1997	0.00357	0.32853	-2.67E-17	0.00
21L05	MW	12/29/1997	0.00387	0.33009	1.48E-13	0.22
21L05	MW	1/20/1998	0.00351	0.33043	1.07E-12	1.60
21L05	MW	1/27/1998	0.00357	0.33064	1.69E-12	2.53
21L05	MW	2/10/1998	0.00357	0.33126	3.57E-12	5.34
21L05	MW	2/17/1998	0.00359	0.33267	7.71E-12	11.54
21L05	MW	3/24/1998	0.00360	0.33681	2.44E-11	36.45
21L05	MW	4/8/1998	0.00362	0.33672	2.40E-11	35.89
21L05	MW	10/29/1998	0.00354	0.33473	1.60E-11	23.93
21L05	MW	1/12/1999	0.00354	0.33152	5.57E-12	8.33
21L06	MW	12/2/1997	0.00363	0.33008	-7.08E-14	-0.11
21L06	MW	12/29/1997	0.00363	0.33013	2.61E-13	0.39
21L06	MW	1/6/1998	0.00349	0.33049	1.22E-12	1.82
21L06	MW	1/20/1998	0.00353	0.33098	2.65E-12	3.96
21L06	MW	1/27/1998	0.00360	0.33106	2.92E-12	4.37
21L06	MW	2/10/1998	0.00356	0.33067	1.92E-12	2.87

(continued)

Table 4.4 (continued)

Sample location/ description*		Sample date	$^{124}Xe/^{132}Xe$	$^{136}Xe/^{132}Xe$	^{136}Xe (cc (STP)/ g H_2O)	^{136}Xe C/C_0 (%)
21L06	MW	2/17/1998	0.00357	0.33056	1.58E-12	2.36
21L06	MW	2/25/1998	0.00360	0.32943	1.23E-12	1.85
21L06	MW	7/14/1998	0.00355	0.33014	-7.41E-14	-0.11
21L07	MW	12/29/1997	0.00358	0.33255	7.55E-12	11.29
21L07	MW	1/20/1998	0.00356	0.33503	1.45E-11	21.61
21L07	MW	1/27/1998	0.00355	0.33335	1.01E-11	15.09
21L07	MW	2/17/1998	0.00359	0.33119	3.57E-12	5.34
21L07	MW	10/29/1998	0.00354	0.33050	1.15E-12	1.72
21L08	MW	12/29/1997	0.00363	0.34535	5.23E-11	78.29
21L08	MW	1/6/1998	0.00352	0.34963	6.69E-11	100.00
21L08	MW	1/20/1998	0.00356	0.33338	1.02E-11	15.25
21L08	MW	2/10/1998	0.00356	0.34046	3.35E-11	50.09
21L08	MW	2/17/1998	0.00355	0.34052	3.27E-11	48.92
21L08	MW	2/25/1998	0.00357	0.34081	3.82E-11	57.17
21L08	MW	3/10/1998	0.00355	0.33362	1.55E-11	23.21
21L08	MW	4/21/1998	0.00355	0.33790	2.30E-11	34.44
21L08	MW	5/12/1998	0.00354	0.34312	3.18E-11	47.59
27D08	MW	1/6/1998	0.00357	0.32837	-2.50E-17	0.00
27D08	MW	6/2/1998	0.00353	0.33016	-1.99E-14	-0.03
PT-1	PW	4/28/1998	0.00355	0.33024	5.01E-13	0.75
PT-1	PW	6/2/1998	0.00362	0.33045	1.08E-12	1.62
PT-1	PW	6/30/1998	0.00358	0.33021	1.43E-13	0.21
PT-1	PW	7/7/1998	0.00359	0.33023	1.61E-13	0.24
PT-1	PW	7/21/1998	0.00359	0.33015	-6.75E-14	-0.10
PT-1	PW	8/11/1998	0.00356	0.33026	2.40E-13	0.36
PT-1	PW	8/25/1998	0.00355	0.33026	2.42E-13	0.36
PT-2	PW	4/28/1998	0.00358	0.33024	1.88E-13	0.28
PT-2	PW	5/26/1998	0.00356	0.32995	-4.52E-13	-0.68
PT-3	PW	2/17/1998	0.00355	0.33013	2.61E-13	0.39
PT-3	PW	3/31/1998	0.00359	0.32920	5.53E-13	0.83
PT-3	PW	4/28/1998	0.00359	0.33021	3.12E-13	0.47
PT-3	PW	6/23/1998	0.00365	0.32945	-1.18E-12	-1.76
PT-3	PW	6/30/1998	0.00368	0.33024	1.91E-13	0.29
PT-3	PW	7/7/1998	0.00371	0.33022	1.43E-13	0.21
PT-3	PW	7/14/1998	0.00369	0.33029	3.74E-13	0.56
PT-3	PW	8/25/1998	0.00372	0.33018	2.02E-13	0.30
PT-3	PW	12/16/1998	0.00355	0.33067	1.92E-12	2.87
PT-4	PW	7/7/1998	0.00369	0.33016	2.71E-14	0.04
PT-4	PW	7/7/1998	0.00370	0.33020	9.00E-14	0.13
PT-4	PW	8/18/1998	0.00365	0.33019	2.15E-13	0.32
PT-4	PW	9/2/1998	0.00361	0.33021	2.60E-13	0.39

(continued)

Table 4.4 (continued)

Sample location/ description*		Sample date	$^{124}Xe/^{132}Xe$	$^{136}Xe/^{132}Xe$	^{136}Xe (cc (STP)/ g H_2O)	^{136}Xe C/C_0 (%)
PT-4	PW	11/23/1998	0.00360	0.33054	1.18E-12	1.76
PT-4	PW	12/22/1998	0.00358	0.33059	1.25E-12	1.87
PT-5	PW	6/30/1998	0.00363	0.33054	1.09E-12	1.63
PT-5	PW	7/7/1998	0.00367	0.33037	5.77E-13	0.86
PT-5	PW	8/18/1998	0.00368	0.33055	9.98E-13	1.49
PT-5	PW	9/2/1998	0.00369	0.33043	8.40E-13	1.26
PT-5	PW	12/1/1998	0.00354	0.33101	2.46E-12	3.68
PT-6	PW	6/9/1998	0.00355	0.33099	2.52E-12	3.76
PT-6	PW	6/23/1998	0.00361	0.33056	1.71E-12	2.56
PT-6	PW	6/30/1998	0.00356	0.33059	1.18E-12	1.76
PT-6	PW	7/7/1998	0.00356	0.33052	1.00E-12	1.50
PT-6	PW	7/21/1998	0.00357	0.33054	1.04E-12	1.55
PT-6	PW	8/18/1998	0.00358	0.33055	1.18E-12	1.76
PT-6	PW	11/24/1998	0.00353	0.33098	2.35E-12	3.51
PT-7	PW	2/17/1998	0.00354	0.33138	3.93E-12	5.88
PT-7	PW	2/24/1998	0.00358	0.33033	4.01E-12	6.00
PT-7	PW	3/10/1998	0.00355	0.33149	7.58E-12	11.33
PT-7	PW	3/31/1998	0.00357	0.33076	5.47E-12	8.18
PT-7	PW	4/14/1998	0.00353	0.33022	4.10E-12	6.13
PT-7	PW	5/19/1998	0.00357	0.33199	5.30E-12	7.93
PT-7	PW	6/9/1998	0.00356	0.33147	3.75E-12	5.60
PT-7	PW	6/30/1998	0.00354	0.33112	2.68E-12	4.01
PT-7	PW	7/7/1998	0.00355	0.33084	2.19E-12	3.27
PT-7	PW	8/3/1998	0.00357	0.33099	2.40E-12	3.59
PT-7	PW	9/2/1998	0.00360	0.33049	1.01E-12	1.51
PT-7	PW	1/12/1999	0.00354	0.33100	2.50E-12	3.74
PT-8	PW	7/14/1998	0.00356	0.32954	-1.32E-12	-1.97
PT-8	PW	7/21/1998	0.00356	0.33017	-5.18E-15	-0.01
PT-8	PW	8/25/1998	0.00359	0.33029	4.74E-13	0.71
Mowry 1	PW	8/3/1998	0.00355	0.33036	6.64E-13	0.99
Mowry 7	PW	3/10/1998	0.00360	0.32914	3.72E-13	0.56
Mowry 7	PW	8/3/1998	0.00356	0.33067	1.55E-12	2.32

*SW= surface water, MW= monitoring well, PW= production well

Notes:

All sampling locations are in the AHF Sub-basin, except Mowry wells 1 and 7, located in BHF Sub-basin.

C/C_o values greater than or equal to 1% consitute a positive detection of tracer.

the aquifer. For ^{124}Xe, the choice of C_0 is based on the maximum point on the 'curve' for the shallow pond samples (Figure 4.15). This choice is based on the assumption that recharge to the aquifer is through the shallow zone of the pond. A value of $1.6x10^{-11}$ cc (STP)/g H_2O is the interpolated maximum of the curve for the shallow samples for ^{124}Xe in Bunting Pond, and is the

Table 4.5

^{124}Xe in Bunting Pond and AHF wells

Sample location/ description*		Sample date	$^{124}Xe/^{132}Xe$	$^{136}Xe/^{132}Xe$	^{124}Xe (cc (STP)/g H_2O)	^{124}Xe C/C_0 (%)
Bunting-East at 4'	SW	1/26/1998	0.00692	0.33007	1.19E-11	74.4%
Bunting-East at 10'	SW	1/13/1998	0.01133	0.33022	2.81E-11	175.6%
Bunting-East at 20'	SW	12/29/1997	0.00853	0.32922	1.76E-11	110.0%
Bunting-East at 20'	SW	1/26/1998	0.01981	0.33009	5.72E-11	357.5%
Bunting-Middle at 4'	SW	1/26/1998	0.00586	0.33013	7.99E-12	49.9%
Bunting-Middle at 5'	SW	1/13/1998	0.00781	0.33026	1.51E-11	94.4%
Bunting-Middle at 9'	SW	1/26/1998	0.00628	0.33014	9.61E-12	60.1%
Bunting-Middle at 15'	SW	12/23/1997	0.00661	0.32867	1.08E-11	67.5%
Bunting-Middle at 15'	SW	1/13/1998	0.01544	0.33033	4.24E-11	265.0%
Bunting-Middle at 20'	SW	1/6/1998	0.01214	0.33028	3.02E-11	188.8%
Bunting-West at 20'	SW	12/29/1997	0.00958	0.32961	2.13E-11	133.1%
21J03	MW	12/16/1997	0.00361	0.32893	1.09E-13	0.7%
21J03	MW	12/29/1997	0.00364	0.33004	2.11E-13	1.3%
21J03	MW	1/6/1998	0.00361	0.33034	1.27E-13	0.8%
21J03	MW	1/20/1998	0.00357	0.33001	9.15E-15	0.1%
21J03	MW	3/3/1998	0.00388	0.32910	8.93E-13	5.6%
21J03	MW	3/31/1998	0.00396	0.32914	1.16E-12	7.2%
21J03	MW	9/1/1998	0.00376	0.33013	5.88E-13	3.7%
21J03	MW	9/9/1998	0.00379	0.33020	6.35E-13	4.0%
21J03	MW	9/15/1998	0.00378	0.33015	6.30E-13	3.9%
21J03	MW	9/22/1998	0.00377	0.33019	5.94E-13	3.7%
21K05	MW	1/20/1998	0.00387	0.33001	8.97E-13	5.6%
21K05	MW	1/27/1998	0.00409	0.33013	1.62E-12	10.1%
21K05	MW	2/10/1998	0.00453	0.33016	3.05E-12	19.1%
21K05	MW	3/10/1998	0.00471	0.32906	3.71E-12	23.2%
21K05	MW	4/21/1998	0.00472	0.33021	3.47E-12	21.7%
21K05	MW	9/22/1998	0.00374	0.33019	4.50E-13	2.8%
21K05	MW	1/27/1999	0.00362	0.33031	1.74E-13	1.1%
21J27	MW	2/24/1998	0.00535	0.32899	5.79E-12	36.2%
21J27	MW	3/3/1998	0.00531	0.32906	5.87E-12	36.7%
21J27	MW	3/31/1998	0.00485	0.32906	4.31E-12	26.9%
21J27	MW	4/7/1998	0.00480	0.32901	4.10E-12	25.6%
21J27	MW	5/5/1998	0.00424	0.32996	2.35E-12	14.7%
21J27	MW	6/16/1998	0.00390	0.33022	9.44E-13	5.9%
21J27	MW	7/28/1998	0.00373	0.33018	4.59E-13	2.9%
21J27	MW	9/22/1998	0.00363	0.33014	2.15E-13	1.3%
21J27	MW	10/6/1998	0.00362	0.33042	1.74E-13	1.1%
21J27	MW	12/1/1998	0.00358	0.33038	1.01E-13	0.6%
PT-1	PW	4/28/1998	0.00355	0.33024	-4.91E-15	0.0%
PT-1	PW	6/2/1998	0.00362	0.33045	1.91E-13	1.2%
PT-1	PW	6/30/1998	0.00358	0.33021	1.19E-13	0.7%

(continued)

Table 4.5 (continued)

Sample location/ description*		Sample date	$^{124}Xe/^{132}Xe$	$^{136}Xe/^{132}Xe$	^{124}Xe (cc (STP)/g H_2O)	^{124}Xe C/C_0 (%)
PT-1	PW	7/7/1998	0.00359	0.33023	1.34E-13	0.8%
PT-1	PW	7/21/1998	0.00359	0.33015	1.56E-13	1.0%
PT-1	PW	8/11/1998	0.00356	0.33026	3.20E-14	0.2%
PT-1	PW	8/25/1998	0.00355	0.33026	3.21E-14	0.2%
PT-2	PW	4/28/1998	0.00358	0.33024	1.07E-13	0.7%
PT-2	PW	5/26/1998	0.00356	0.32995	1.69E-14	0.1%
PT-3	PW	2/17/1998	0.00355	0.33013	-5.36E-14	-0.3%
PT-3	PW	3/31/1998	0.00359	0.32920	1.35E-14	0.1%
PT-3	PW	4/28/1998	0.00359	0.33021	9.00E-14	0.6%
PT-3	PW	6/23/1998	0.00365	0.32945	1.85E-13	1.2%
PT-3	PW	6/30/1998	0.00368	0.33024	4.05E-13	2.5%
PT-3	PW	7/7/1998	0.00371	0.33022	4.87E-13	3.0%
PT-3	PW	7/14/1998	0.00369	0.33029	4.39E-13	2.7%
PT-3	PW	8/25/1998	0.00372	0.33018	4.74E-13	3.0%
PT-3	PW	12/16/1998	0.00355	0.33067	1.70E-14	0.1%
PT-4	PW	7/7/1998	0.00369	0.33016	4.12E-13	2.6%
PT-4	PW	7/7/1998	0.00370	0.33020	4.39E-13	2.7%
PT-4	PW	8/18/1998	0.00365	0.33019	2.66E-13	1.7%
PT-4	PW	9/2/1998	0.00361	0.33021	1.49E-13	0.9%
PT-4	PW	11/23/1998	0.00360	0.33054	1.35E-13	0.8%
PT-4	PW	12/22/1998	0.00358	0.33059	1.25E-13	0.8%
PT-5	PW	6/30/1998	0.00363	0.33054	2.50E-13	1.6%
PT-5	PW	7/7/1998	0.00367	0.33037	3.66E-13	2.3%
PT-5	PW	8/18/1998	0.00368	0.33055	3.53E-13	2.2%
PT-5	PW	9/2/1998	0.00369	0.33043	3.82E-13	2.4%
PT-5	PW	12/1/1998	0.00354	0.33101	2.81E-14	0.2%
PT-6	PW	6/9/1998	0.00355	0.33099	-8.57E-15	-0.1%
PT-6	PW	6/23/1998	0.00361	0.33056	2.24E-13	1.4%
PT-6	PW	6/30/1998	0.00356	0.33059	4.99E-14	0.3%
PT-6	PW	7/7/1998	0.00356	0.33052	6.66E-14	0.4%
PT-6	PW	7/21/1998	0.00357	0.33054	7.47E-14	0.5%
PT-6	PW	8/18/1998	0.00358	0.33055	7.90E-14	0.5%
PT-6	PW	11/24/1998	0.00353	0.33098	-4.09E-14	-0.3%
PT-7	PW	2/17/1998	0.00354	0.33138	-8.23E-14	-0.5%
PT-7	PW	2/24/1998	0.00358	0.33033	-2.39E-14	-0.1%
PT-7	PW	3/10/1998	0.00355	0.33149	-1.17E-13	-0.7%
PT-7	PW	3/31/1998	0.00357	0.33076	-5.19E-14	-0.3%
PT-7	PW	4/14/1998	0.00353	0.33022	0.00E+00	0.0%
PT-7	PW	5/19/1998	0.00357	0.33199	6.26E-14	0.4%
PT-7	PW	6/9/1998	0.00356	0.33147	2.56E-14	0.2%
PT-7	PW	6/30/1998	0.00354	0.33112	-2.83E-16	0.0%
PT-7	PW	7/7/1998	0.00355	0.33084	2.89E-14	0.2%
PT-7	PW	8/3/1998	0.00357	0.33099	4.16E-14	0.3%

(continued)

Table 4.5 (continued)

Sample location/ description*		Sample date	$^{124}Xe/^{132}Xe$	$^{136}Xe/^{132}Xe$	^{124}Xe (cc (STP)/g H_2O)	^{124}Xe C/C_0 (%)
PT-7	PW	9/2/1998	0.00360	0.33049	1.24E-13	0.8%
PT-7	PW	1/12/1999	0.00354	0.33100	2.00E-15	0.0%
PT-8	PW	7/14/1998	0.00356	0.32954	4.55E-14	0.3%
PT-8	PW	7/21/1998	0.00356	0.33017	4.90E-14	0.3%
PT-8	PW	8/25/1998	0.00359	0.33029	1.14E-13	0.7%
Mowry 1	PW	8/3/1998	0.00355	0.33036	8.85E-15	0.1%
Mowry 7	PW	3/10/1998	0.00360	0.32914	1.83E-14	0.1%
Mowry 7	PW	8/3/1998	0.00356	0.33067	1.57E-14	0.1%

*SW= surface water, MW= monitoring well, PW= production well

Notes:

All sampling locations are in the AHF Sub-basin, except Mowry wells 1 and 7, located in BHF Sub-basin.

C/C_o values greater than or equal to 1% consitute a positive detection of tracer.

value used to calculate tracer percentages at wells. The detection limit for both tracers corresponds to a C/C_0 value of 1 percent. This detection limit is not the laboratory detection limit for the isotope, but is instead the level of the tracer isotope that the investigators could attribute with confidence to being in excess of the pre-dosing background level in the aquifers. The amount of tracer corresponding to a C/C_o value greater than 1 percent is referred to herein as "excess" tracer. The date that a C/C_o value of 1 percent was attained in the well was taken as the arrival time of the leading edge of tracer at the well. Because a sample from a given well was typically not collected on the very day the value of C/C_o actually reached 1 percent, arrival times were interpolated or extrapolated by plotting a breakthrough curve. Breakthrough curves for tracers in production and monitoring wells are included on upcoming pages in this section.

Surface Water

During the dosing period, the water columns in both ponds were profiled for temperature and pH. The results indicate an absence of a thermocline (Table 4.6).

Thorough lateral mixing in both ponds was evidenced by the laterally uniform distribution of tracer concentrations, as indicated by samples from collected from similar depths and on the same day (Figure 4.15). Lateral mixing is the result of wind-driven circulation.

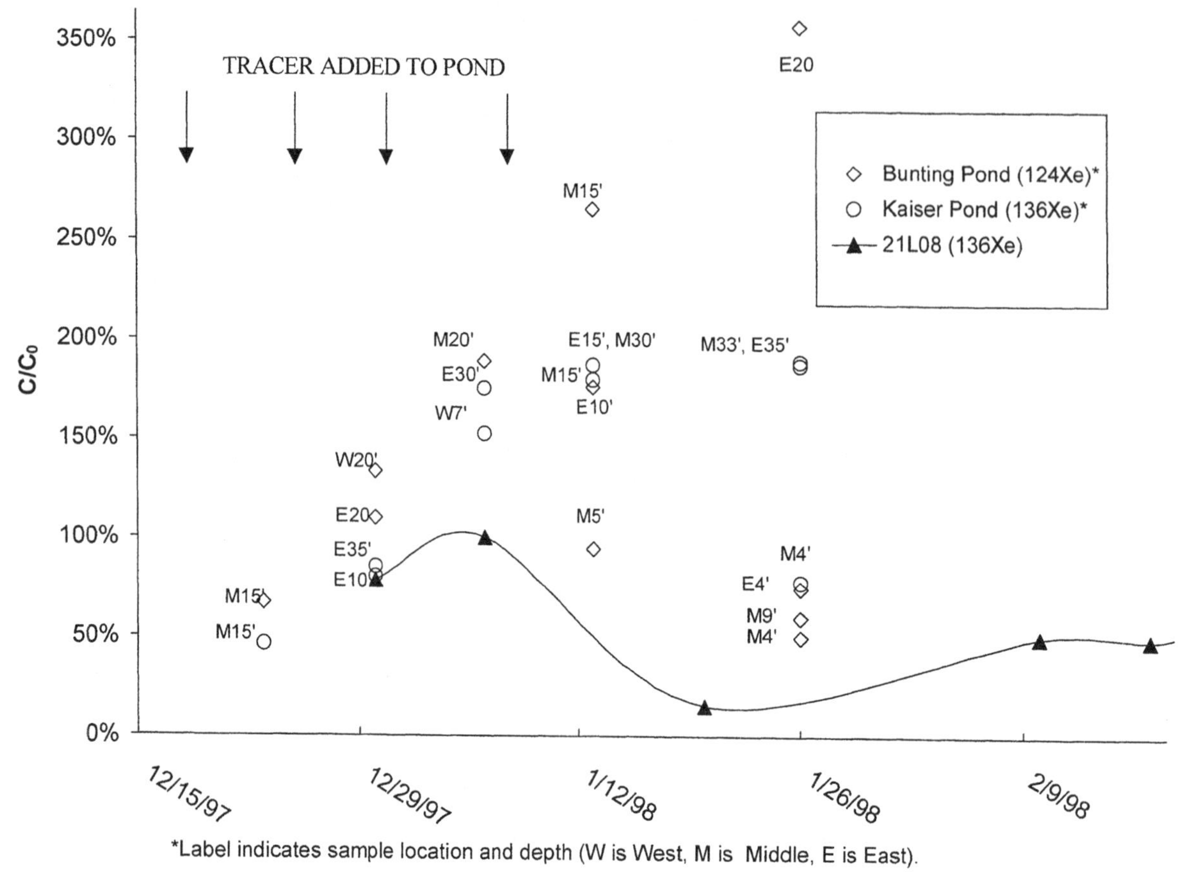

Figure 4.15 Tracer detections in Kaiser Pond and Bunting Pond, and breakthrough curve for monitoring well 21L08. Kaiser 'M' sample location is same as Kaiser 'S' sample location on Figure 3.1

Table 4.6

Field parameters measured in Kaiser and Bunting Ponds (AHF) on 12/30/97

Depth (feet)	Temperature (°C)	Conductivity (umohs/cm)	Dissolved O_2 (mg/l)	pH	ORP* (mV)
Kaiser Pond (depth to bottom at lowest point = 71 feet)					
3	10.7	760	9.4	7.9	423
16	10.5	759	9.3	7.8	423
33	10.4	759	9.2	7.8	423
54	10.3	760	9.1	7.8	424
Bunting Pond (depth to bottom at lowest point = 38 feet)					
3	8.9	767	11.1	7.9	368
16	8.9	767	10.9	7.9	367
33	8.9	767	10.7	7.9	367

*ORP = oxidation reduction potential

Surprisingly, deeper samples had much higher tracer concentrations than shallow samples. In fact, deep samples from Bunting Pond continued to increase throughout the monitoring period, while deep samples from Kaiser pond leveled off at a high concentration during the last two sampling dates. Theoretical considerations predict that the lack of a thermocline would result in a constant depth profile, despite the fact the tracers were added at substantial depths in the ponds. Solubility of a gas (the tracer is a noble gas isotope) in water is temperature dependent, and the concentration of tracer in the ponds should change with time only as a result of recharge and diffusion (loss) across the water/air interface.

With reference to the analysis of Kaiser Pond seepage (see previous section), active recharge is believed to occur almost exclusively through the upper sides of the ponds and not through the bottom, and this may have had an effect on the vertical distribution of the tracer in the ponds. In order to flux out of the pond and into groundwater, tracers would have had to migrate upward in the water column from the injection depth. Despite the lack of evidence of a thermocline based on temperature profiles, vertical mixing in the ponds was incomplete. Release of tracers from the pond to groundwater may have been controlled by slow vertical diffusion across an unseen interface between shallow and deep layers in the water column. Although monitoring of pond water did not continue beyond two weeks, the long mild-sloped tails on the breakthough curves in down-gradient wells suggest a long tracer residence time in the ponds.

Groundwater

Tracer breakthrough curves for the monitoring wells provide further evidence that recharge occurs throughout the sides of the ponds. Well 21L08, just adjacent to Kaiser Pond on the down-gradient side, had the highest tracer level of any well. The level of ^{136}Xe measured in 21L08 matched the level measured in the shallow section of Kaiser pond (Figure 4.15) in late December 1997. The arrival time following dosing was almost immediate: a high concentration was observed just one week after tracer introduction, and the maximum tracer concentration was observed just two weeks after the initial tracer application.

Breakthrough curves for ^{136}Xe tracer in other wells (down-gradient of Kaiser Pond) are provided on Figure 4.16. Comparing the maximum tracer level observed in 21L08 to ^{136}Xe levels observed in other monitoring and production wells, substantial mixing of non-tagged

(a)

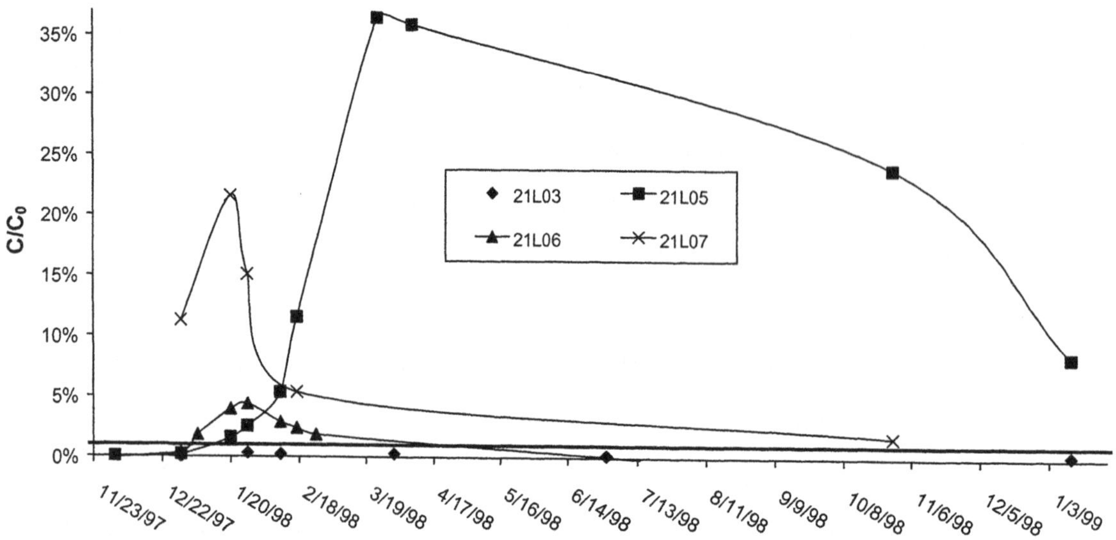

(b)

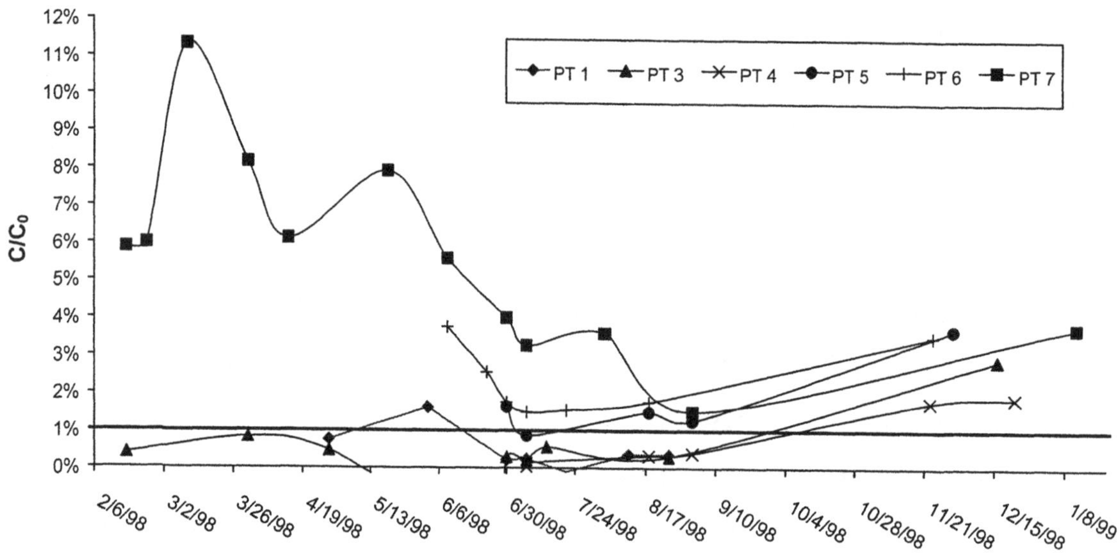

Figure 4.16 ^{136}Xe tracer (introduced in Kaiser Pond) in AHF portion of the study area: (a) monitoring wells and (b) Peralta-Tyson production wells. A C/C_0 value equal to or greater than 1% constitutes a positive tracer detection.

water with water tagged with tracer from Kaiser Pond was observed. The relatively large overall depths and long screens of the wells may have resulted in dilution of the tracer during purging, which in turn, may have lead to low observed tracer concentrations. Water that was recharged before or after the tagged water moved through the basin may have also diluted the tracer signal. The complete absence of tracer in well 21L03 is somewhat surprising, given its location between the pond and the production wellfield. However, the top of the screen in 21L03 is at an elevation that is typically 40 feet below the phreatic surface, and tagged water may have migrated along a more shallow flow-path. In addition, the much greater depth of this well, compared to other monitoring wells, may have resulted in dilution of tagged water with water from deeper zones, resulting in a tracer concentration that was below the detection limit. It is also possible that the well is in a 'dead zone' – an area through which tracer did not migrate. However, the boring log for this well does not indicate low-permeable materials, other than a thin silt zone at the 115-foot depth below ground surface.

Breakthrough curves for ^{124}Xe in groundwater wells down-gradient of Bunting Pond are provided in Figure 4.17. Similar to wells monitoring the migration of ^{136}Xe down-gradient of Kaisier Pond, a substantial dilution of ^{124}Xe tracer was evident in wells down-gradient of Bunting Pond. The level of ^{124}Xe tracer fell below the detection limit in 21K05, 21J27, PT4, and PT5 after October 1998. Among monitoring wells close to Bunting Pond, well 21J27 had a breakthrough curve with a significantly higher maximum concentration and a shorter tail. These characteristics are attributed to the well's relatively shallow screened interval. Groundwater movement is believed to be more rapid in the shallower depths of the aquifer.

Neither tracer was detected in monitoring well 27D08, which was monitored occasionally as the distal limit of the tracer study (Tables 4.4 and 4.5). In addition, tracers were not detected in production wells PT-2 and PT-8. A relatively small amount of water was produced from these wells compared to other PT wells; tracer may have been detected had a greater amount of water been pumped from them. In addition, it should be noted that PT-2 was monitored only on two occasions in the spring of 1998, prior to positive detections of ^{136}Xe tracer in PT-1 and PT-3 and ^{124}Xe tracer in PT-3 and PT-6. Tracer may have been detected eventually in PT-2 if it had been monitored into the summer of 1998. PT-8 was monitored on three occasions in the summer of 1998, when either one or both of the xenon tracers had already been detected in other wells (except PT-2). The absence of both xenon tracers in this well may be a consequence of this well

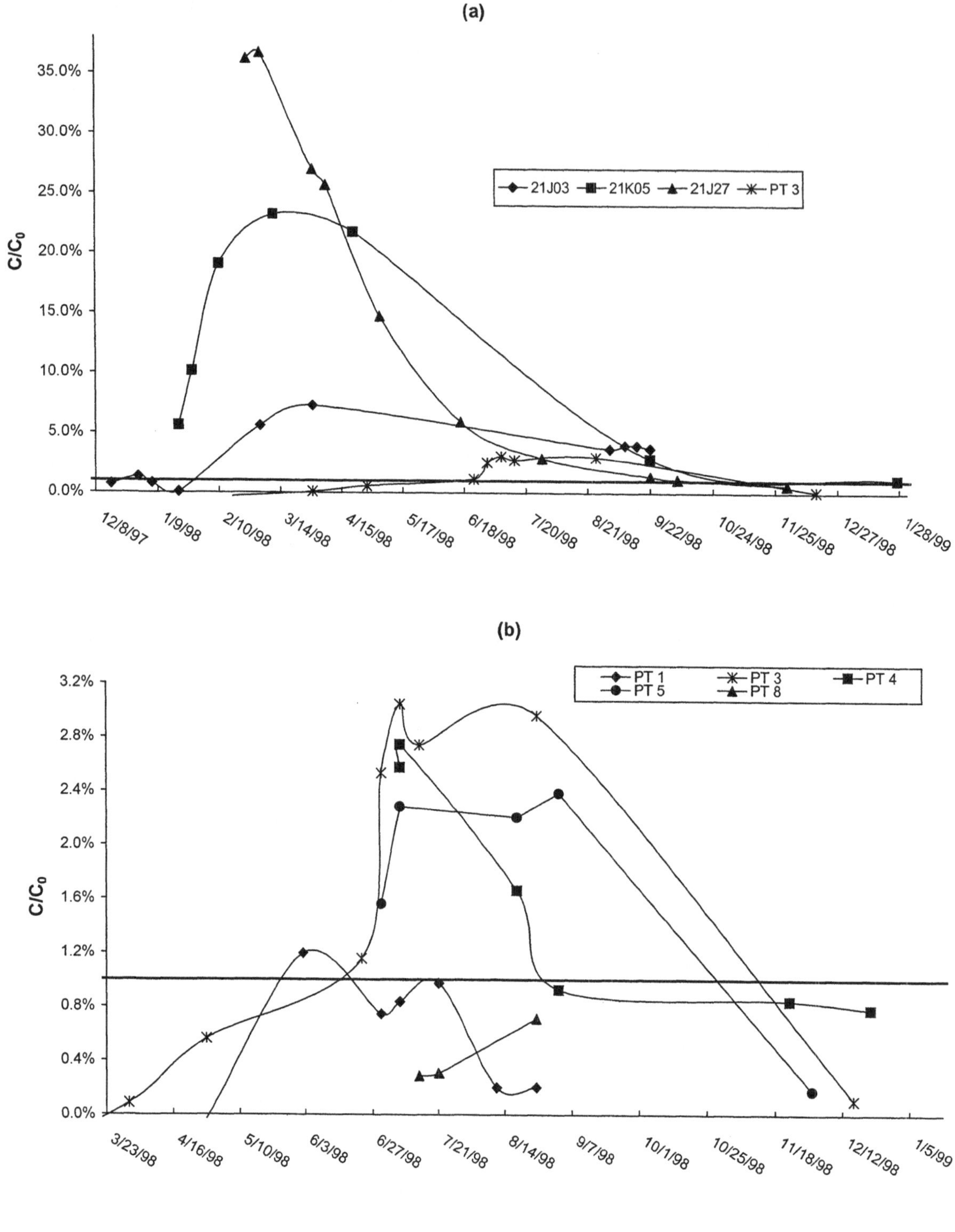

Figure 4.17 ^{124}Xe tracer (introduced in Bunting Pond) in AHF portion of study area: (a) monitoring wells (and PT-3) and (b) PT production wells. A C/C_o value equal to or greater than 1% constitutes a positive tracer detection.

being screened exclusively in the zone beneath the fine-grained soil layer, suggesting that tracer migrated preferentially in the zone above this layer. No other production wells have perforated intervals located exclusively below this layer. The absence of tracers in PT-8 is noteworthy but, admittedly, more samples from the well collected over the full one-year monitoring period would have provided a basis for a stronger argument. Even better would have been sampling of deep groundwater from monitoring wells located up-gradient of the wellfield; however, no such monitoring wells were available for this project (21L03 was not screened exclusively in the deep zone).

Most of the breakthrough curves for the various monitoring wells near Bunting and Kaiser Ponds have pre-peak slopes that rise suddenly, and post-peak tails that decline gradually. The gradual decrease is attributed in part to a long residence time of tracer in the ponds and a corresponding gradual rate of tracer flux to groundwater. The authors sought to compare the declining tail of 21L05's breakthrough curve to that of the water column in Kaiser Pond. Unfortunately, sufficient pond data were not available because surface water monitoring did not continue for a long enough time period after dosing.

The arrival times of tracers in monitoring wells near the ponds were generally the same. However, not all wells near the ponds had breakthrough curves with long, gradual tails. Furthermore, substantial differences in peak concentrations were observed. The difference in the curves among wells 21L05, 21L06, and 21L07 initially came as a surprise, since all three wells are close to Kaiser Pond and are screened in approximately the same depth interval. Relative to the curve for 21L05, 21L07's curve had a very sharp decline. The decline of tracer concentration in 21L06 was also rather rapid. The peak concentration in 21L06, having reached only 5 percent, was one seventh of the peak in 21L05. These differences could be due to spatial variability in recharge rate along the walls of Kaiser Pond. Another possible explanation is that wells 21L06 and 21L07 may intercept water that seeps from a shallow embayment in the southeast corner of the pond. The large diversion pipe that delivers water from Alameda Creek to the pond empties into this embayment, and inflow of water may have rapidly flushed tracer from the embayment after dosing. Continued inflows probably prevented re-equilibration of water quality between the embayment and the rest of the pond. Consequently, water seeping from the southeast corner of the pond likely had a weak tracer signal with a sharp decline, as observed in 21L06 and 21L07. Wells 21J03 and 21K05, down-gradient of Bunting Pond, have approximately the same screened interval; however, 21J03's breakthrough curve has a much lower maximum and a longer tail. These differences are likely due to spatial variability in recharge along the south wall of Bunting Pond.

As shown in Figures 4.16 (b) and 4.17 (b), breakthrough curves for most of the production wells had multiple peaks. Most probably, the tracer elution curve for PT-7 has at least four maxima, including a return to levels well above the detection limit after falling to a level near the detection limit. Interestingly, the concentration in PT-7 hit a second maximum in May when the well was not pumping (Table 4.2a), only to drop again in the summer of 1998 with resumption of heavy pumping of PT-7 and other PT wells. The small third maximum doesn't seem to fit the pattern, but the concentration in PT-7 increased as demand on PT-7 and other PT wells dropped in the fall and early winter of 1998. A comparison between PT-7 tracer concentration and total wellfield demand is provided on Figure 4.18. The existence of multiple peaks may be due to complicated pumping patterns at the wellfield. As total wellfield demand increases, as it does in the late spring and summer (refer to Table 4.2a and Figure 4.18), multiple high capacity pumping wells in close proximity will inevitably interfere with each other hydraulically. It is plausible that the interference is greater in the shallower zone (above the fine-grained soil layer), because the wells fully penetrate the shallow zone but not the full thickness of the overall aquifer all the way down to bedrock (refer to Figure 2.6). Consequently, the fraction of inflow from the deep zone may increase with total demand. If the tracer concentration were lower in the deeper zone, then the tracer concentration in the well would be diluted with increased inflow from the deep zone.

Duration of pumping prior to sample collection should probably also be considered among the possible reasons for the observed rise and fall pattern of the breakthrough curves. It may take a certain amount of time after the pump is turned on for the deeper zones to contribute water to the well.

A more robust interpretation of the tracer results at the wellfield could be offered through three-dimensional modeling of flow and transport within the study area. The spatial resolution of the model would likely need to be sufficient to simulate hydraulic interference effects among individual pumping wells. The model would also need to have the capability to simulate the effect resulting from partially penetrating wells drawing from more than one water-bearing zone.

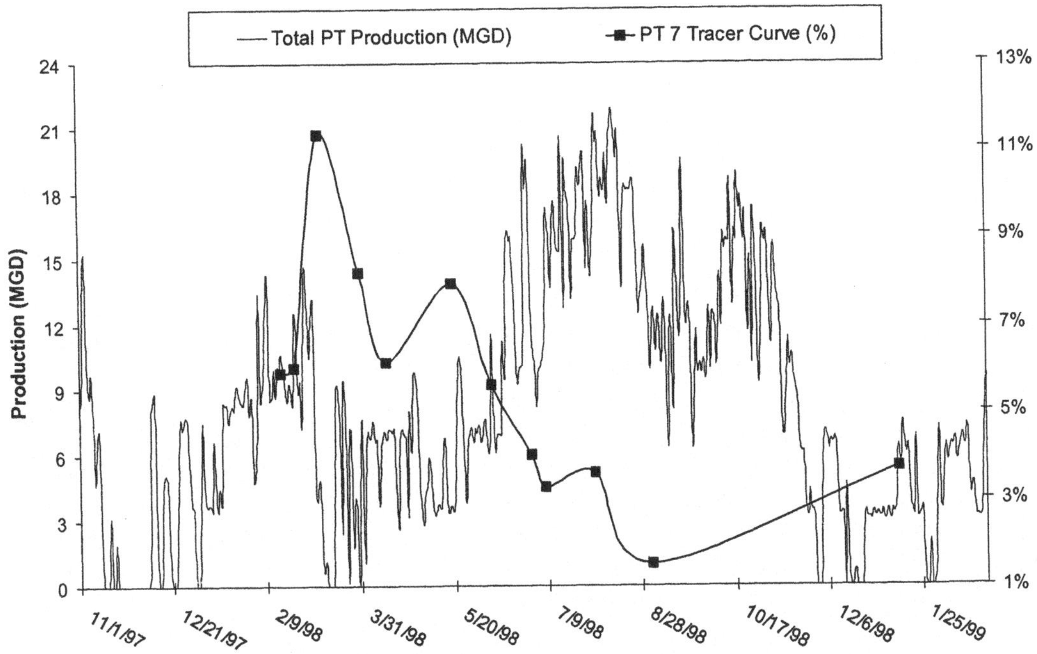

Figure 4.18 Daily total PT Wellfield production and ^{136}Xe tracer in PT-7 (AHF)

Tracer travel rates to monitoring and production wells, based on times coinciding with the earliest detection and the peak concentration, are calculated in Tables 4.7 and 4.8. Arrival times based on first detection are also presented on Figures 4.19a and 4.19b. As a consequence of production schedules and pump maintenance, some production wells were not available for sampling for several months in the winter and spring of 1998. Therefore, exact arrival times could not be determined for all wells. The times given in Tables 4.7 and 4.8 and Figures 4.19a and 4.19b are calculated from the first day of tracer input to the time of first arrival at the well. For wells sampled and analyzed frequently, the initial part of the tracer curve is interpolated, as described previously, to give the arrival time. For each well that was sampled or analyzed infrequently, a "less-than-or-equal to" arrival time is given since there were too few measurements to interpolate more specific values.

The times between dosing and the first detection of tracer in the above-mentioned wells are quite short. For example, ^{136}Xe tracer was observed in PT-7 (approximately 1,200 ft from Kaiser Pond) only two months after the date of the first introduction of tracer into Kaiser Pond. Because this initial sample had a C/C_0 value of 6%, and the value for a positive detection is 1%, it can be concluded that the velocity of the leading edge of the tracer mass is greater than 7,000

Table 4.7

Estimated travel rates of ^{136}Xe (added to Kaiser Pond) based on time to first detection and peak concentration.

Calculation of ^{136}Xe tracer velocity according to first detection						
*Well	Distance to Kaiser Pond (ft)	Initial dose date Kaiser Pond	First detection of tracer	First arrival (days)	Tracer travel rate (ft/day)	Tracer travel rate (ft/yr)
PT-3	1,600	12/16/97	3/31/98	105	15.2	5,600
PT-4	1,780	12/16/97	10/15/98	303	5.9	2,100
PT-5	1,400	12/16/97	Before 6/30/98	< 196	> 7.1	> 2,600
PT-6	1,300	12/16/97	Before 6/9/98	< 175	> 7.4	> 2,700
PT-7	1,200	12/16/97	Before 2/17/98	< 63	> 19.0	> 7,000
Mowry1	2,700	12/16/97	8/3/98	230	11.7	4,300
Mowry7	2,700	12/16/97	8/3/98	< 230	> 11.7	4,300
21L05	210	12/16/97	1/14/98	29	7.2	2,600
21L06	190	12/16/97	1/6/98	21	9.0	3,300
21L07	80	12/16/97	Before 12/29/97	< 13	> 6.2	> 2,200
21L08	40	12/16/97	Before 12/29/97	< 13	> 3.1	> 1,100
Calculation of ^{136}Xe tracer velocity according to detection of peak concentration						
*Well	Distance to Kaiser Pond (ft.)	Est'd date of peak in g.w. at pond	Date of peak at well	Arrival time of peak (days)	Tracer travel rate (ft/day)	Tracer travel rate (ft/yr)
PT-3	1,600	1/6/98	12/16/98	344	4.7	1,700
PT-4	1,780	1/6/98	12/22/98	350	5.1	1,900
PT-5	1,400	1/6/98	12/1/98	329	4.3	1,600
PT-6	1,300	1/6/98	6/9/98	154	8.4	3,100
PT-7	1,200	1/6/98	3/10/98	63	19.0	7,000
Mowry1	2,700	1/6/98	unknown	--	--	--
Mowry7	2,700	1/6/98	unknown	--	--	--
21L05	210	1/6/98	Before 3/24/98	< 77	> 2.7	> 1,000
21L06	190	1/6/98	1/27/98	21	9.0	3,300
21L07	80	1/6/98	1/20/98	14	5.7	2,100
21L08	40	1/6/98	1/6/98	0	--	--

*All wells are located in the AHF Sub-basin, except Mowry wells 1 and 7, located in BHF Sub-basin

Abbreviations/ symbols: est'd= estimated; g.w.= groundwater; -- = not available

feet per year. Such a high velocity indicates a high hydraulic conductivity for AHF sediments within the study area, given the low gradient prevalent at the time of the tracer study. The time to peak in PT-7 is particularly noteworthy, not only because of the fast travel rate (7,000 feet/year) but also because it occurred prior to the peak concentration in up-gradient monitoring well 21L05. Evidently, a preferential pathway exists, possibly located immediately west of 21L05 and 21L03, extending from Kaiser Pond to the PT Wellfield. It is plausible that this preferential pathway is some way associated with the Hayward Fault.

Table 4.8

Estimated travel rates of ^{124}Xe tracer (added to Bunting Pond) in AHF study area, based on time to first detection and peak concentration

Calculation of ^{124}Xe tracer velocity according to first detection						
Well	Distance to Kaiser Pond (ft)	Initial dose date Bunting Pond	First detection of tracer	First arrival (days)	Tracer travel rate (ft/day)	Tracer travel rate (ft/yr)
PT-1	1,920	12/16/97	Before 6/2/98	< 168	> 11.4	> 4,200
PT-3	1,810	12/16/97	6/23/98	189	9.6	3,500
PT-4	1,830	12/16/97	Before 7/7/98	< 203	> 9.0	> 3,300
PT-5	1,770	12/16/97	Before 6/30/98	< 196	> 9.0	> 3,300
21J03	50	12/16/97	12/29/97	13	3.8	1,400
21K05	30	12/16/97	1/15/98	30	1.0	400
21J27	530	12/16/97	Before 2/24/98	< 70	> 7.6	> 2,800
Calculation of ^{124}Xe tracer velocity according to detection of peak concentration						
Well	Distance to Bunting Pond (ft)	Est'd date of peak in g.w. at pond	Date of peak at well	Arrival time of peak (days)	Tracer travel rate (ft/day)	Tracer travel rate (ft/yr)
PT-1	1,920	1/13/98	6/2/98	140	13.7	5,000
PT-3	1,810	1/13/98	7/7/98	175	10.3	3,800
PT-4	1,830	1/13/98	7/7/98	175	10.5	3,800
PT-5	1,770	1/13/98	9/2/98	232	7.6	2,800
21J03	50	1/13/98	3/31/98	77	0.6	200
21K05	30	1/13/98	3/10/98	56	0.5	200
21J27	530	1/13/98	3/3/98	49	10.8	3,900

Abbreviations/ symbols: est'd= estimated; g.w.= groundwater; -- = not available

Theoretically, the time to reach peak concentration should be better than the time of first detection as an indicator of the advective velocity of the tracer. The effects of longitudinal dispersion along the specific pathways followed by the tracer within this relatively short monitoring period should not effect the progress of the tracer center of mass (i.e., the time to reach peak concentration in a down-gradient well). The reference start time to compute advective velocity by this method should be the estimated time that the concentrations of tracers at the pond/aquifer interface were at their peak; i.e., when C/C_0 in groundwater was 100%. Accordingly, the reference times for ^{136}Xe (added to Kaiser Pond) and ^{124}Xe (added to Bunting Pond) tracers were taken to be 1/6/98 and 1/13/98, respectively. These reference dates are consistent with the method followed to determine reference C/C_0 values of 100%, as discussed previously in this chapter. The lower half of Tables 4.7 and 4.8 were developed according to this methodology. Admittedly, there is some uncertainty involved with these peak concentration start dates. However, the initial dose date is a known value. Therefore, travel time and advective

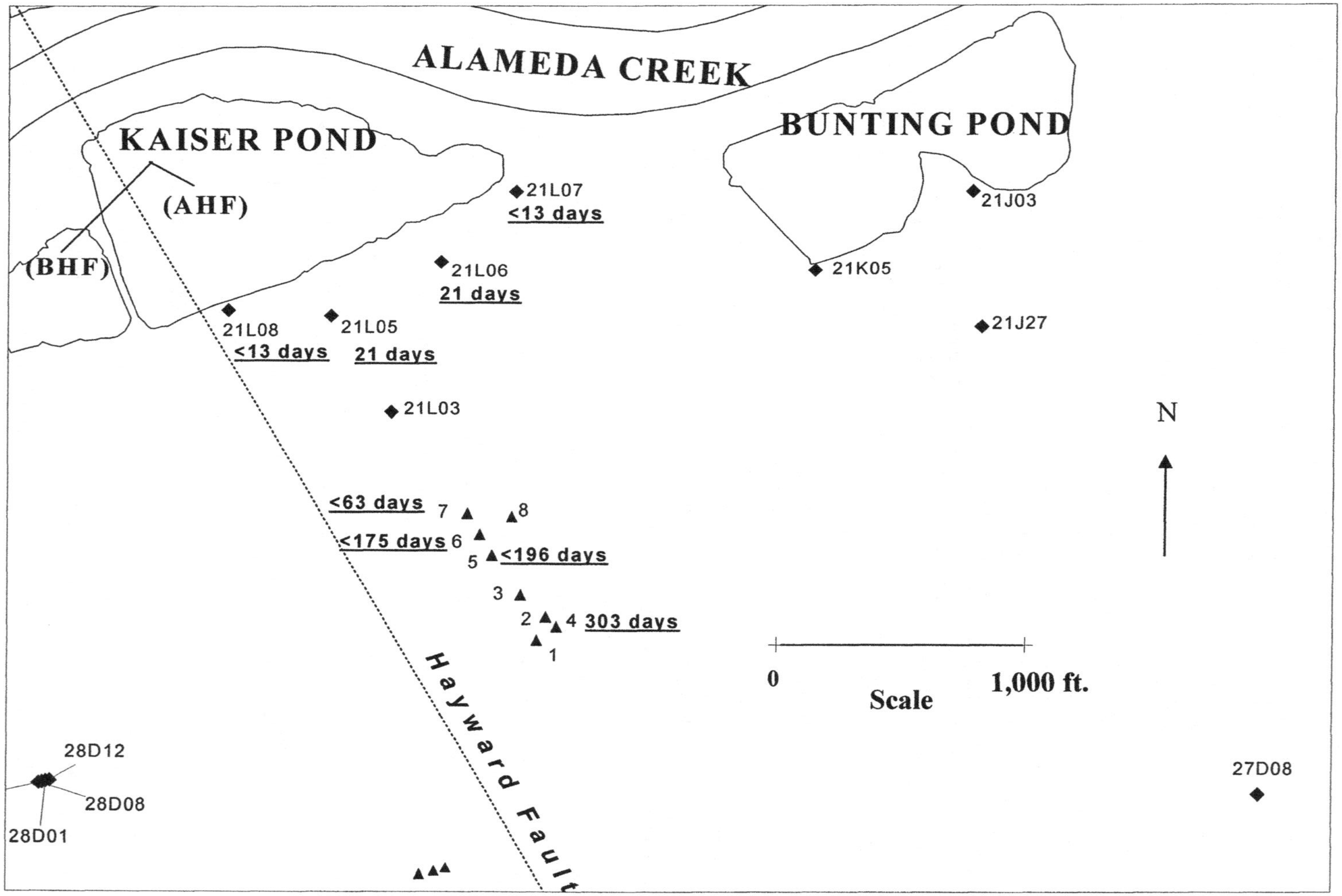

Figure 4.19(a) Estimated arrival times for the leading edge of ^{136}Xe tracer (dosed in Kaiser Pond) in AHF monitoring wells and PT (AHF) production wells.

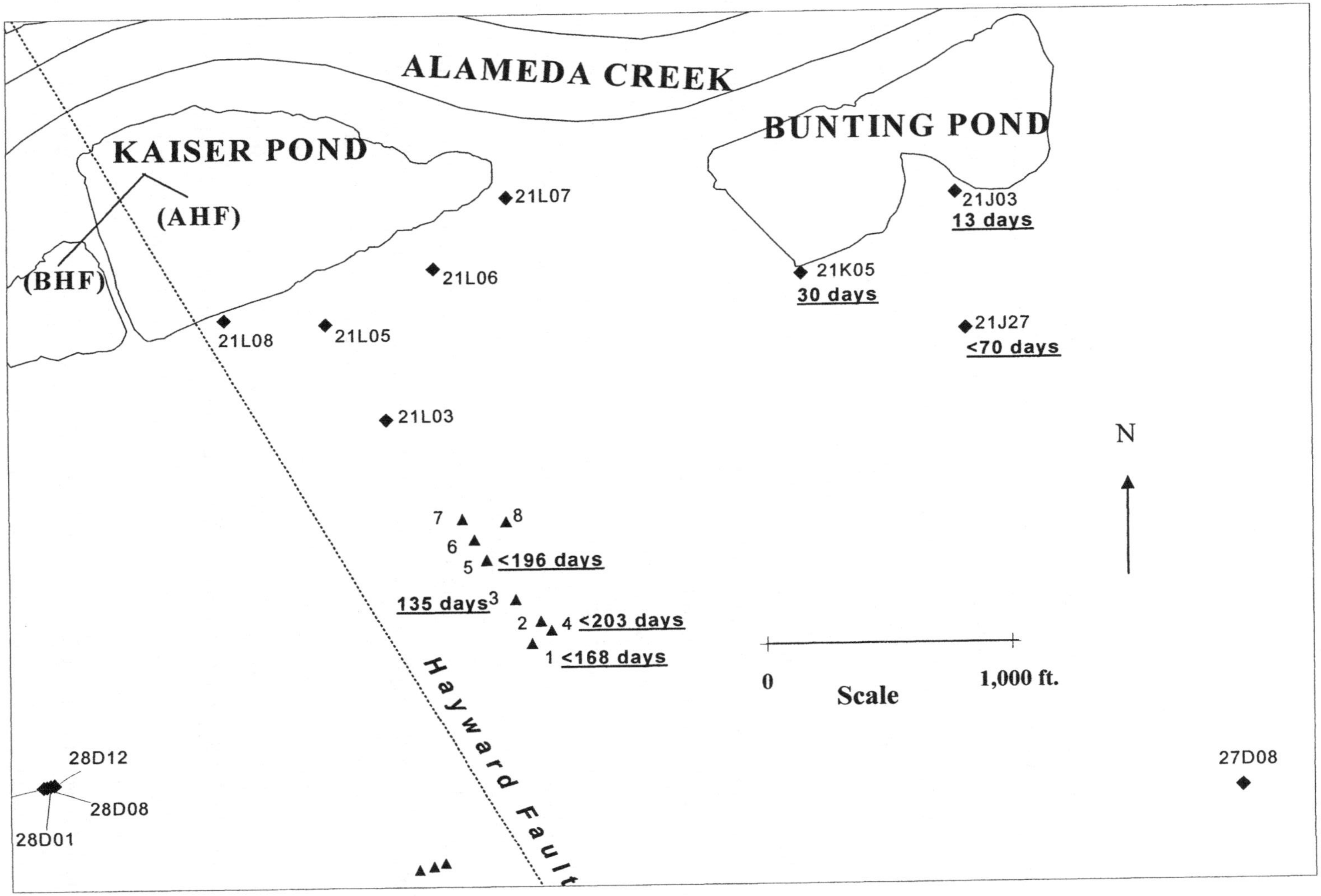

Figure 4.19(b) Estimated arrival times for the leading edge of ^{124}Xe tracer (dosed in Bunting Pond) in AHF monitoring wells and PT (AHF) production wells

velocity are discussed hereinafter in terms of the time to first detection, as reported in the upper half of Tables 4.7 and 4.8. Considering the steep ascent to the peak concentration on the breakthrough curves for most of the wells, the error from possible local longitudinal dispersion is probably small enough to justify calculating advection velocity from time to first detection.

It is useful to compare the observed times of arrival with the value calculated from measured gradients and a pump-test determined transmissivity of 3,800,000 gpd/ft for the AHF Aquifer (DWR 1968). Assuming a saturated thickness of 450 feet (based the estimated depth to bedrock of 500 feet), the average hydraulic conductivity over the full thickness of the aquifer should be roughly 1,200 ft/day. Using the gradients on Table 4.3 and an assumed porosity of 0.3, the depth-averaged advective travel time should be no less than 900 days. This value is approximately 14 times the estimate given by the time for ^{136}Xe to peak in PT-7 and 6 times the estimate given by the time for ^{124}Xe to peak in PT-1. The Darcy calculation may be flawed because PT-8 was used in the gradient calculation. Because PT-8 is screened exclusively in the zone beneath the fine-grained soil layer, whereas the other wells are screened in both zones, the water level in PT-8 may not have represented head in the upper interval of the aquifer through which the tracers may have preferentially migrated.

The absence of detections in PT-8 and the large differences between the Darcy calculation and the tracer arrival times are consistent with the notion that tracer migrated more quickly in the shallow part of the aquifer above the fine-grained layer. Preferential travel in shallow intervals of the aquifer is also evidenced by the relatively high concentration observed in well 21J27 (screened over a shallow interval), the fast arrival and substantial tracer signal at PT-7 (perforations in this well begin at shallower depth relative to most other PT wells), and the observed lack of tracer in 21L03 (screened at a deeper interval relative to other monitoring wells). In this unconfined aquifer, a fast pathway may in fact exist at the water table.

Groundwater flow modeling (outside the scope of this study) would help quantify the extent to which hydraulic conductivity differences are responsible for the apparent slower movement of groundwater in the lower part of the aquifer, below the fine-grained soil layer. Geometric effects may also be contributing to the variation in flow velocity with depth. One potential reason for possible slower velocities in the deep groundwater is the greater thickness of the lower zone. The perforations penetrate the upper zone by 70 percent but penetrate the lower zone by only (an estimated) 25 percent. The total depth of the production wells is 200 feet, and the depth to bedrock is estimated to be 500 feet. Streamlines along the bottom of the aquifer would probably be longer and slower, even without a lower hydraulic conductivity. The condition of the perforations could also tend to favor flow from shallower groundwater, although ACWD has not conducted any spinner tests to determine the distribution of inflow with depth in the wells.

The rapid arrival of tracer relative to the probable depth-averaged flow velocity (the latter determined by age-dating or the Darcy calculation) gives the appearance of substantial longitudinal dispersivity from the hypothetical view of the aquifer as a homogeneous, single layered system that extends from ground surface to bedrock. But the slope of the pre-peak climbing side of the breakthrough curves can give a sense of local longitudinal dispersivity along the specific pathways followed by the tracer within the time frame of the experiment. The estimate would likely err on the high side because tracer was added to the ponds over a four-week period. Modeling would be advantageous in quantifying dispersion coefficients, since the shape of the breakthrough curve in wells should be dependent not only on dispersivity but also on the duration of the dosing period.

The post-peak falling side of the curves cannot be considered for estimates of local longitudinal dispersivity, as the gradual slope could be a consequence of non-local dispersive effects such as the slow release of tracer from the ponds. Tracer persistence in the production wells could also be attributed in part to the contribution from slower streamlines that follow a less direct, more circuitous path from the ponds, or from streamlines that lead to the well from deep zones within the aquifer.

In addition to comparing peak concentrations in order to characterize the degree of attenuation of tracer with distance down-gradient, it is interesting to contrast the areas under the breakthrough curves for the various wells with the area under Kaiser Pond's breakthrough curve. Using a fourth-order polynomial to fit tracer curves, the area under 21L08's curve is 60 percent of the area under the Kaiser Pond's curve, and the area under PT-7 is 19 and 31 percent of the area under the curves for Kaiser Pond and well 21L08, respectively. Since the mass flux of water into the wells is not considered, the amount of tracer mass recovered cannot be calculated by simple comparison of curve area. However, rough mass balances can be calculated to give an indication of the degree of mixing and dilution that occurs in the AHF sub-basin. If it can be assumed that pumping at the PT Wellfield is the only sink for water (output), then the tracer concentrations measured in PT wells (times the total volume pumped) can be compared with the known mass of tracer that was introduced into the ponds. In this analysis, loss of tracer to the atmosphere, and loss of water tagged with tracer to distal and deeper parts of the basin, were not considered. This one-dimensional approach is useful only for gross comparisons of mass fluxes. The total tracer input of ^{136}Xe into Kaiser Pond was 1,040 cc (STP), and the volume of ^{124}Xe added to Bunting Pond was 124 cc (STP).

If the total mass of tracer introduced were evenly distributed throughout the volume of the pond at a single point in time with no dilutions or losses, the concentration of excess ^{136}Xe in Kaiser would have been 7.80 x 10^{-10} cc (STP)/g H_2O. The actual maximum observed concentration in the pond was 15 percent of that concentration. This lower concentration is believed to be a consequence of losses to the atmosphere, loss to the BHF sub-basin via flow to BHF portion of Kaiser Pond, and slow dispersion of the tracer over time (which limited the amount of tracer detected at the sampling locations). (As noted above, there is no evidence for subsurface leakage across the Hayward Fault.) If all of the tracer that was put into Kaiser Pond were to have been found in the 1,400 acre-ft that pumped from PT-7 over the period of the study, the average concentration in PT-7 would have been 6.02 x 10^{-10} cc (STP)/g H_2O. The observed maximum concentration in PT-7, however, was only 1 percent of that number. If all of the Kaiser tracer were to have been found in the water pumped from all of the PT wells over the course of the study (approximately 10,000 acre-ft), then the average concentration in the pumped water would have been 8.42 x 10^{-11} cc (STP)/g H_2O. The maximum observed concentration at the PT Wellfield (at PT-7) was about one order of magnitude less than that. Therefore, it can be concluded that a large fraction of the water recharged from Kaiser Pond does not reach the PT Wellfield within one year.

Similarly, without considering atmospheric losses of ^{124}Xe tracer added to Bunting Pond, the volume-weighted average concentration of excess ^{124}Xe in this pond should have been 3.18 x 10^{-10} cc (STP)/g H_2O. However, the maximum measured ^{124}Xe concentration in Bunting Pond was only 18 percent of that value. If all of the tracer mass added to Bunting Pond were to have been recovered in the 582 acre-ft pumped from PT-3 during the course of the study, the average concentration in PT-3 would have been 1.73 x 10^{-10} cc (STP)/g H_2O. But the maximum measured concentration in PT-3 was only 0.3 percent of that value. If all the tracer mass were to be recovered in the 10,000 acre-feet pumped from all the PT wells in summation, the average concentration at the wellfield would have been 1.01 x 10^{-11} cc (STP)/g H_2O. However, the highest concentration observed at the wellfield was only 5 percent of this value. The extremely small amount of mass recovery in PT wells during the study period further supports the notion that a small fraction of the tracer-laden water traveled quickly, probably through a shallow, high conductivity layer, while the bulk of the tagged water moved more slowly through deeper intervals.

A more thorough accounting of the amount of tagged water that did not reach the wellfield must consider losses due to volatilization. Any excess xenon in water in contact with the atmosphere or soil gas can volatilize. Chapra and Reckhow (1983) present equations to estimate lake surface air/water mass transfer coefficients as a function of wind speed and molecular weight of the volatile constituent. Even if the mass transfer coefficient were as low as 15 cm/day (corresponding to an average wind speed of 5 miles per hour), volatilization would constitute 60 percent of the loss of tracer from Kaiser Pond, with the remaining 40 percent fluxed to the aquifer. This estimate is based on a pond area of 24 acres and an average seepage rate of 9 acre-feet per day during the monitoring period. The estimate assumes that seepage occurs from shallow depths in the pond, such that the concentration at the air/water interface was assumed to be approximately the same as that in water seeping from the pond.

Significant additional volatilization may also occur during seepage if flow from the pond were unsaturated. Despite the presence of soil solids, a high contact area between per volume of seepage water may arise with unsaturated flow. Once in the saturated zone, the volatilization rate should be greatly reduced because of the absense of turbulent effects and the likely presence of a capillary fringe that would tend to greatly increase resistance to gas-water exchange. At production wells, high entrance velocities can increase chances for turbulence, and hence, volatilization. Volatilization is particularly of concern if the pumping water level is drawn down below the top of perforations, potentially leading to entrained air in the water discharged by the well. Had air been entrained in water at ACWD's production wells, it would have been observed in the form of 'excess air' in the dissolved gas samples containerized in copper tubes. No excess air was detected. It does not appear that tagged water for this study underwent significant volatilization at any point following seepage from the ponds.

The above discussion does underscore the nature of noble gas tracers: through inert, they are not conservative in surface water or during vadose zone transport.

Modeling to Help Interpret Tracer Results

It is apparent that complexity of the system, the potential for volatilization losses, and limited number of observations and monitoring wells have presented challenges in interpreting the results and quantifying flow and transport parameters. Though outside the scope of this study, development of a combined pond and groundwater model could help overcome some of these challenges. Flow and transport parameters could be quantified through calibrating the model to the observed tracer results and water levels.

ALAMEDA CREEK (AHF) TRACER STUDY (^{129}Xe)

In this experiment, Alameda Creek (upstream of Rubber Dam No. 3) was dosed with ^{129}Xe tracer. The results, provided in Table 4.9, are not expressed in terms of C/C_0 because an estimate of a value for C_0 (the peak concentration of tracer in seepage water) would have been almost completely arbitrary. A reasonable estimate of C_0 was precluded by the brief residence time in the creek, the difficulty in ensuring a uniform concentration in surface water along the dosed segment of the creek, and a lack of wells to monitor creek seepage exclusively. Therefore, the authors have opted to quantify concentrations of ^{129}Xe in terms of the 'del' notation (introduced in Chapter 3) in lieu of the C/C_0 parameter.

For samples collected on 9/14/98 and 9/15/98, a δ^{129}Xe value greater than 1 $^{o}/_{oo}$ constituted a positive detection of tracer. For samples collected on all other dates, a positive detection of ^{129}Xe tracer corresponded to a δ^{129}Xe of 2 $^{o}/_{oo}$ or greater. Dosing the creek was successful, with a δ^{129}Xe of 99 $^{o}/_{oo}$ in surface water recorded on 8/27/1998. The creek study was different from the pond studies in that a smaller volume of water was tagged with tracer, and tagged water may have been lost via surface water diversions to the ponds and flow to downstream segments of Alameda Creek. This includes tracer movement from the creek to Bunting Pond two weeks after the last tracer application, when the gate was opened between the creek and the pond. There is a possibility, therefore, that the tracer detected in 21K05 and in 21J03 was recharged through Bunting Pond rather than through the creek. Moreover, a large pipeline conveys water ponded ahead of Rubber Dam No. 3 to a point in the creek downstream of the dam. ACWD's recharge operation precluded closing off the pipeline during the tracer

Table 4.9

Results of Alameda Creek tracer (^{129}Xe) experiment (AHF)

Sample location/description*		Date collected	$\delta^{129}Xe$ (‰)	Sample location/description*		Date collected	$\delta^{129}Xe$ (‰)
Alameda Creek	SW	8/25/98	**2.94**	21K05	MW	9/22/98	**3.29**
Alameda Creek	SW	8/27/98	**99.65**	21K05	MW	12/29/98	**9.78**
				21K05	MW	1/27/99	0.33
21H11	MW	8/24/98	0.74				
21H11	MW	9/15/98	**1.03**	21J27	MW	9/22/98	0.92
21H11	MW	10/6/98	**2.04**	21J27	MW	10/6/98	0.36
				21J27	MW	12/1/98	0.34
21L05	MW	10/29/98	-0.62	21J27	MW	3/10/99	**6.07**
21L05	MW	1/12/99	0.08				
21L05	MW		**4.73**	PT-3	PW	8/25/98	0.96
				PT-3	PW	12/16/98	0.66
21J03	MW	9/1/98	0.65				
21J03	MW	9/9/98	**1.11**	PT-7	PW	8/3/98	0.91
21J03	MW	9/15/98	0.83	PT-7	PW	9/2/98	0.80
21J03	MW	9/22/98	**1.14**	PT-7	PW	1/12/99	-0.36
21J03	MW	12/29/98	**2.70**				

*SW= surface water, MW= monitoring well, PW= production well
Notes:
Numbers in bold represent positive detections of tracer.
Experiment was conducted exclusively in AHF portion of the study area; all samples were collected from AHF locations
Alameda Creek samples were collected upstream of Rubber Dam No. 3.

experiment, and consequently, a significant amount of tracer mass was likely flushed downstream through this pipeline.

Tracer was first detected in wells 21H11 and 21K05 (see Figure 3.1 for locations of these wells) approximately one month after the initial dosing. Specifically, well 21H11 had a slight amount of tracer on 9/15/98, with a $\delta^{129}Xe$ of 1.03 ‰. Well 21K05 had a larger amount of tracer, with a $\delta^{129}Xe$ of 3.3‰ on 9/22/98. By 12/29/98, the level of tracer in 21K05 had climbed to a $\delta^{129}Xe$ value 9.8‰; however, tracer was not detected in this well on 1/27/99 ($\delta^{129}Xe$ was below 1‰). Tracer was detected in well 21J03 on 12/29/98 ($\delta^{129}Xe$ was 2.7‰) but was not observed on 9/1, 9/9, 9/15 or 9/22/98. Since no tracer was present in well 21H11 on 8/24/98, it can be concluded that the arrival time of tracer in 21H11 was between 8/24/98 and 9/15/98. The observation of no detectable level of tracer in well 21H11 on 10/6/98 is consistent with the contention that only a small amount of tagged creek water recharged the aquifer. For well 21K05, the calculated minimum travel rate for the leading edge of the tracer is over 15,000 ft/yr.

This value, which is too high to be attributed to groundwater transport, supports the hypothesis that the ^{129}Xe tracer was transferred to Bunting Pond via the surface water connection prior to being received in well 21K05.

On balance, the number of detections of the ^{129}Xe tracer was rather limited. The authors believe that only a small volume of the tagged water seeped to the aquifer before surface water transfers and losses occurred. Dilution of the tagged water by water from other sources resulted in tracer levels dropping below the detection limit in areas distant from the recharge source. There were no positive detections of the tracer at the PT Wellfield. In addition to unintended surface water transfers and losses, a lack of available monitoring wells in the northeasterly part of the study area also limited the success of the creek tracer experiment. Only one well, 21H11, was available for monitoring at this location.

BHF POND TRACER (^{22}Ne) STUDY

The results of the BHF tracer experiment are provided in Table 4.10. For this experiment, a large volume of water (approximately 4,000 acre-ft in Stevenson Pond and Lago los Osos) was tagged with ^{22}Ne tracer. Results are given in terms of the del notation, and a positive detection of tracer corresponds to a $\delta^{22}Ne$ value of $0.8^{o}/_{oo}$ or greater. Arrival times for the tracer are plotted in Figure 4.20.

Surface Water

Surface water samples reveal that processes similar to those described for AHF ponds may also occur at Lago los Osos. Four surface water samples indicated excess ^{22}Ne. Two samples collected on the same day from the Pit S portion of Lago los Osos had very different concentrations, with the deeper sample having a much higher concentration. Unlike Kaiser and Bunting Ponds, however, a thermocline was evident in the depth-profile of temperature, pH, and oxidation-reduction potential, as indicated in Table 4.11. In addition to incomplete vertical mixing, horizontal mixing may have also been incomplete. Two samples from Lago los Osos, collected only two days apart, at similar depths but at different locations, had very different ^{22}Ne concentrations. As discussed in Chapter 2, Lago los Osos was constructed from former Pits R and S. One of the samples was collected from former Pit R and the other from Pit S (see

Table 4.10

Results of BHF tracer (^{22}Ne) experiment

Sample location/description*		Date collected	δ ^{22}Ne (‰)	Sample location/description*		Date collected	δ ^{22}Ne (‰)
Pit S, Northeast, 30'	SW	8/25/98	**3.09**	28D11	MW	12/8/98	-0.47
Pit S, Northest, 5'	SW	8/27/98	**21.43**	28D11	MW	3/10/99	0.11
Pit S, South, 35'	SW	8/27/98	**474.45**	28D11	MW	5/25/99	-0.35
Pit H, 5'	SW	9/2/98	**2.21**	28D12	MW	11/24/98	1.05
Rainbow Lake	SW	9/2/98	0.25	Mowry 1	PW	10/14/98	0.56
20J04	MW	9/29/98	0.43	Mowry 1	PW	12/9/98	0.40
20J04	MW	10/20/98	0.26	Mowry 1	PW	4/22/99	0.38
20J04	MW	11/9/98	0.40	Mowry 1	PW	6/8/99	0.21
20J06	MW	1/5/99	-0.19	Mowry 6	PW	12/16/98	-0.74
20R03	MW	10/13/98	**4.27**	Mowry 7	PW	12/9/98	-0.42
20R03	MW	11/9/98	**1.20**	Mowry 7	PW	1/12/99	-0.27
20R03	MW	1/5/99	0.71	Mowry 7	PW	2/29/99	0.49
20R04	MW	11/17/98	**3.11**	Mowry 7	PW	6/15/99	0.07
20R04	MW	1/5/99	**3.54**	PT-7	PW	4/27/99	0.01
20R07	MW	1/27/99	-0.23				
20R07	MW	4/6/99	-0.12				
20R07	MW	6/1/99	-0.47				

*SW= surface water, MW= monitoring well. Surface water sample descriptions include depth of sample.

Notes:

Numbers in bold represent positive detections of tracer.

Experiment was conducted exclusively in BHF portion of the study area; all samples were collected from BHF locations, except PT-7, located in AHF Sub-basin.

locations "NE" and "S", respectively, on Figure 3.2). Incomplete mixing is not surprising because the levees separating the former pits still exist below elevation 0 feet MSL.

Two "downstream" ponds, which receive water from the Lago los Osos directly or indirectly, were sampled to determine the extent of tracer distribution throughout the Quarry Pits complex. A sample collected from Horseshoe Lake (the Pit H portion thereof) on 9/2/98 had an excess of ^{22}Ne, but a sample from Rainbow Lake on the same day did not.

Groundwater

The following wells were monitored for ^{22}Ne tracer: 20J04, 20J06, 20R03, 20R04, 20R07, 28D11, 28D12, and certain Mowry production wells. Positive tracer detections (Table 4.10) were observed only in 20R03 and 20R04, with tracer detected in 20R03 on 10/13/98 (the maximum for this well) and 11/9/98, and in 20R04 on 11/17/98 and 1/5/99. The concentrations of tracer collected from 20R04 on these two dates were approximately equal to each other. The

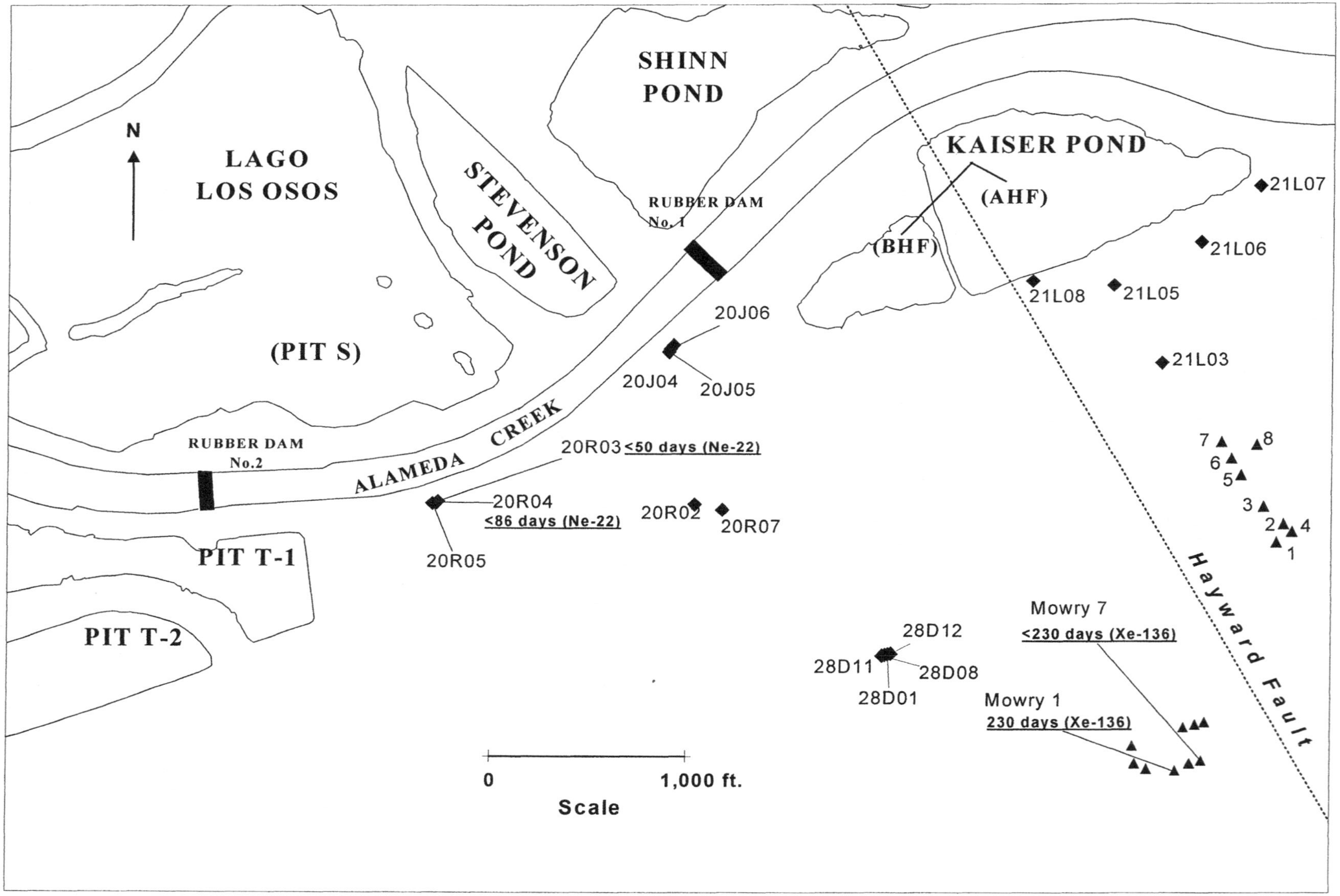

Figure 4.20 Arrival time of tracers in BHF wells

Table 4.11

Field parameters measured in Pit S (BHF) on 8/20/98

Depth (feet)	Temperature (°C)	Conductivity (umohs/cm)	Dissolved O_2 (mg/l)	pH	ORP* (mV)
2	23.1	753	7.7	8.8	311
3	23.1	753	7.6	8.8	310
15	23.0	752	7.1	8.8	310
20	22.9	751	5.7	8.7	307
25	21.9	740	1.5	8.2	316
30	19.0	704	0.2	7.8	105

*Oxidation-reduction potential
Depth to bottom at lowest point in Pit S = 74 feet

cluster that includes 20R03and 20R04 is located immediately adjacent to Alameda Creek, and is in alignment with the Mowry Wellfield and the southerly shoreline of Lago los Osos. Wells 20J04 and 20J06, which are also located adjacent to Alameda Creek, fall on a line between Stevenson Pond and the Mowry wellfield. These two wells, whose screens are at approximately the same depth range as those at 20R03 and 20R04, were monitored over the same time period but had no detections of tracer. Likewise, samples from well 20R07, which is located between Pit S and the Mowry Wellfield, were analyzed on 1/27/99, 4/6/99, and 6/1/99 and had no excess ^{22}Ne. Well 28D12, screened in the deeper zone of the Newark Aquifer, had a very small tracer anomaly on 11/24/98. Since this anomaly could not be confirmed with a duplicate analysis, or with a sample from approximately the same time period, it is not a confirmed positive detection and is not discussed further.

The travel times from Pit S to wells 20R03 and 20R04 were less than or equal to 50 days and 86 days, respectively. The approximate distance between the pit and these wells is 500 feet. The tracer travel rate for the 'leading edge' (~3,700 ft/yr) is therefore in the range of travel rates observed in the AHF sub-basin. The direction of travel from Stevenson Pond and Pit S, as inferred by the tracer results is to the southwest, apparently bypassing the Mowry Wellfield. Piezometric head data in Figures 4.6 and 4.7 do not give an overwhelmingly clear picture of the direction of groundwater flow in the study area. However, a comparison of piezometric heads among the wells with shallowest screens (20J04, 20R03, and 20R07) suggests that Newark Aquifer groundwater mounded around Kaiser Pond-BHF, and that seepage from Pit S and Stevenson Pond flowed to the southwest. The absense of ^{22}Ne tracer in wells other than 20R03 and 20R04 is consistent with the hypothesis that at least some seepage water from Kaiser Pond-

BHF flows directly toward the Mowry Wellfield, while recharge from Pit S and Stevenson Pond either bypasses the wellfield or flows toward the wellfield over a long circuitous route. This flow pattern may, or may not, be maintained under hydrologic conditions different from those of the tracer experiment. Since 20R03 and 20R04 had tracer detections and are located on the opposite side of Alameda Creek from Pit S, it can be concluded that the creek did not act as a hydraulic barrier to the extent that seepage was prevented from crossing under the creek.

If the single datum for a shallow pit sample is used as C_0, then detections of ^{22}Ne tracer in groundwater, expressed as C/C_0, ranged from 5 to 20 percent. For the two pairs of samples from roughly the same dates, well 20R04, which is screened over a deeper interval (74.5 to 84.5 feet b.s.s.) in the Newark Aquifer, had higher tracer levels than well 20R03 (screened between 38 and 58 feet b.g.s.). This suggests that there may be a more transmissive regime in the middle depths of the Newark Aquifer, relative to the overlying interval.

Neon-22 tracer was not present in any of the samples collected from Mowry wells 1 and 7 during the monitoring period up through June 15, 1999. Mowry Wells 1 and 7, which draw from the Newark Aquifer, have the youngest mean water ages and would likely be the first among the production wells to receive water from the recharge facilities. These two wells had an average combined production rate of 214 acre-feet per month between the time of dosing of BHF ponds (August 1998) and the end of monitoring (June 1999). Although no ^{22}Ne tracer was detected in these two wells, ^{136}Xe tracer, introduced into the AHF portion of Kaiser Pond, was detected at 1 percent (as C/C_0) in Mowry 1, and at 2 percent in Mowry 7, in August, 1998 (Table 4.4 and Figure 4.20). Xenon-136 tracer must have entered Kaiser Pond–BHF through the conduit pipe that connects Kaiser Pond-AHF with Kaiser Pond-BHF. Assuming the time for this surface water transfer to occur was negligible, it can be concluded that the minimum groundwater travel time from Kaiser Pond-BHF to the Mowry wells is 8 months over the distance of approximately 2,700 feet. This corresponds to a travel rate of approximately 4,000 ft/yr (see Table 4.7).

The estimated hydraulic conductivity of the Newark Aquifer at the Mowry Wellfield, based on a pump test performed on Mowry 7 in 1996, is 310 feet/day. No monitoring wells exist in direct alignment between Kaiser Pond-BHF and the Mowry Wellfield; therefore, a gradient could not be calculated along this alignment. However, by comparing piezometric heads at 20J04 and Mowry 8, (Figures 4.8 and 4.9), a reasonable estimate is that the gradient between

Kaiser Pond (BHF) and the Mowry Wellfield probably does not exceed 0.005 ft/ft. This would give a Darcy velocity of 6.2 feet per day, assuming a porosity of 0.25. The corresponding travel time from Kaiser Pond (BHF) to the wellfield, without considering longitudinal dispersion, would be approximately 15 months, given a distance of 2,700 feet. Potential reasons why the tracer arrived more quickly than this 15-month estimate include a preferential pathway along the Hayward Fault, non-uniform permeability with depth, and possible leakage across the fault from the AHF side.

Neon-22 tracer was not found in the aquifers below the Newark Aquifer. Wells that were sampled for tracer in the Centerville-Fremont Aquifer included monitoring well 28D11 and Mowry Well 6 (Table 4.10). This absence of tracer in the CF Aquifer could be attributed to the existence of an effective aquitard across the study area, as indicated in Figure 2.7. However, because no Centerville-Fremont monitoring wells are located near Newark Aquifer wells where tracer was indisputably detected, the available results cannot yet be used to characterize leakage potential across the intervening aquitard. There are other observations from the study, however, that provide evidence of a competent aquitard. Wells whose screens are perforated in the Centerville-Fremont and Deep Aquifers had high tritium contents and relatively old ages, suggesting no influence of very recently recharged water in these aquifers. These observations do not preclude the possibility that water from Pit S and Stevenson Pond is ultimately a source of water to Mowry Wells screened in the Centerville-Fremont and Deep Aquifers. Leakage from the Newark Aquifer to the deeper aquifers probably occurs over a large, diffuse area, and the path from the Quarry Pits to the wells may be long and circuitous.

While little can be said about groundwater velocity or changes in water quality parameters based on the ^{22}Ne tracer data, much can be inferred about the character of the flow, based on the lack of tracer observed in these wells. While the Newark aquifer has younger ages than the deeper aquifer layers, it seems that the water tagged with ^{22}Ne tracer was not a significant-or at least not an immediate-source of recharge to the Newark Aquifer in the region between the Quarry Pits complex and the Mowry Wellfield. The appearance of ^{136}Xe in Mowry wells 1 and 7, likely from the small Kaiser-BHF Pond, suggests that this pond could be an important source of recharge for Mowry wells screened in the Newark Aquifer. Leakage across the fault is unlikely a significant source of water for the Mowry Wellfield. Previous regional groundwater modeling in conjunction with a detailed water balance of the AHF and

BHF sub-regions has indicated that leakage across the fault amounts on average to 150 acre-feet per year. This amount represents total annual groundwater flow across the fault integrated over the entire length (of the fault) through the Niles Cone; therefore, the volume of fault leakage within the study area is likely to be much less.

SURFACE WATER AND GROUNDWATER CHEMISTRY

Monitoring water chemistry and tracer concentrations simultaneously provides a means to discern and quantify non-conservative behavior exhibited by various natural constituents while in the process of transport in the subsurface. To achieve the best possible results for this component of the project, it would have been necessary to collect multiple samples over time from each monitoring point, and to develop concentration-versus-time curves that could be compared with the tracer breakthrough curves. Such comparisons would have helped distinguish transformation and/or adsorption-related effects on water quality from advection, dispersion, or dilution. It may have been also necessary to add unique tracers to certain other surface water entities that were not tagged in this study but whose percolation might have reached the wells within the monitoring period (e.g., Alameda Creek between Rubber Dams No. 1 and 3). All such additional recharge sources would also need to have been monitored over time for water chemistry as well as tracer concentrations.

Because of budget and operational constraints, the water chemistry portion of this project could not be carried out to this level of effort. At most wells, only one round of samples for water chemistry analysis was collected, with the sampling date timed to coincide with the arrival of the peak concentration of tracer. In following the tracer center of mass (peak), it was hoped that the same 'unit' of water would be sampled as it migrated from one monitoring point to the next. If the timing were to prove successful and the amount of dilution with non-tagged water demonstrated to be minimal, water quality changes between monitoring points could be attributed to non-conservative processes. Age dating results revealed beforehand that the minimum-dilution criterion would not be met at production wells, but it seemed plausible that sufficiently undiluted tagged-water could be encountered at monitoring wells near the ponds for at least a brief period of time.

However, this hope was not to be realized. At all wells in which tracers were detected, with the possible exception of well 21L08, tracers arrived in very low concentrations, suggesting dilution with non-tagged sources. Dilution was attributed to the combination of long well screens in conjunction with non-uniform concentration of tracer within the interval of aquifer depth intercepted by the well screens. As discussed earlier in the chapter, the aquifer probably received seepage water (and hence tracer) from the dosed ponds near the phreatic surface. The non-tagged component in the water column of each well was probably older water that had seeped from the pond prior to dosing. However, there's some chance it could have also been recharge from non-tagged sources, mainly segments of Alameda Creek between Rubber Dams No. 1 and No. 2 (BHF) and No. 1 and 3 (AHF and BHF).

What was achieved was essentially a water quality snapshot-in-time through an effort that was not able to take advantage of the tracer experiments to the extent necessary to achieve precise quantification of transformation and/or retardation parameters. Despite these limitations, the generated set of data offers strong indications of plausible chemical processes and provides a useful basis for possible future confirmatory studies. Generally, the discussion herein focuses on the overall difference between surface water and nearby groundwater, which was more notable than the relatively minor differences observed among groundwater monitoring points.

The observed significant chemical differences between groundwater and surface water suggest that the clogging layer lining the ponds may impart a beneficial effect on recharge water quality. Documented improvements to groundwater quality during recharge include an elimination of male-specific coliphage, and reductions in TOC, non-point-source-MTBE and nitrate. An increase in TDS (as evidenced by increases in major cations) is the only significant degradation of water quality that may have occurred during recharge. Modeling geo-chemical processes could enhance interpretation of the water quality data. However, geochemical modeling was outside the scope of this project.

The complete set of surface water and groundwater chemistry data can be found in the tables in Appendix B. Field data (dissolved oxygen, conductivity, pH, temperature, and oxidation-reduction potential (ORP)) appear in Table B-1. General water quality parameters, analyzed by contract laboratories, appear in Table B-2. Table B-2 also lists the results of TOC analyses performed by ACWD's in-house laboratory. Results of analyses performed at LLNL are provided in Tables B-3 through B-6. They include ultra low level MTBE and TOC analyses shown in Table B-3 and trace metals data in Tables B-4, B-5, and B-6.

Selected water quality data from the Appendix tables are also presented in various graphs and tables incorporated within this chapter. Each graph and associated table include results of samples collected from a specific pond (Kaiser Pond, Bunting Pond or Pit S) and a set of wells associated with that pond.

Major Cations

Concentrations of major cations in surface water and vicinity wells are indicated in Figures 4.21, 4.22, and 4.23.

With reference to Kaiser Pond, Bunting Pond, and Pit S, concentrations of the primary cations generally increased from surface water to groundwater. The exception to the pattern was the lack of significant difference in magnesium concentrations between Bunting Pond and down-gradient wells. The increases were approximately 65, 11, 50, and 60 percent for potassium, magnesium, sodium, and calcium, respectively. This pattern indicates that dissolution of sedimentary material is taking place until chemical equilibrium is re-established. Dissolution of carbonate minerals and a coupled reaction in which exchange takes place with clays containing Na and K are the likely sources of these constituents for subsurface waters. Vadose zone materials in arid climates frequently contain naturally occurring, easily leachable inorganic constituents. Johnson et al. (1999) found significant increases in the major cations during artificial recharge, which they attribute to dissolution of carbonate minerals. Changes in the concentrations of these species from near-pond wells to production wells are insignificant, indicating that chemical reactions take place on short time and distance scales.

As discussed later in this section, certain other water quality parameters were not consistent with the apparent increase in cation concentrations from Kaiser and Bunting Ponds to groundwater.

Major Anions

The sum of major anions, including nitrate, sulfate, and chloride, did not increase significantly from surface water to groundwater and therefore don not match the trend in cations (Figures 4.24, 4.25, and 4.26). This is an apparent contradiction, since charge balance must be maintained. Bicarbonate, which was not included in the bar graphs, probably makes up for the

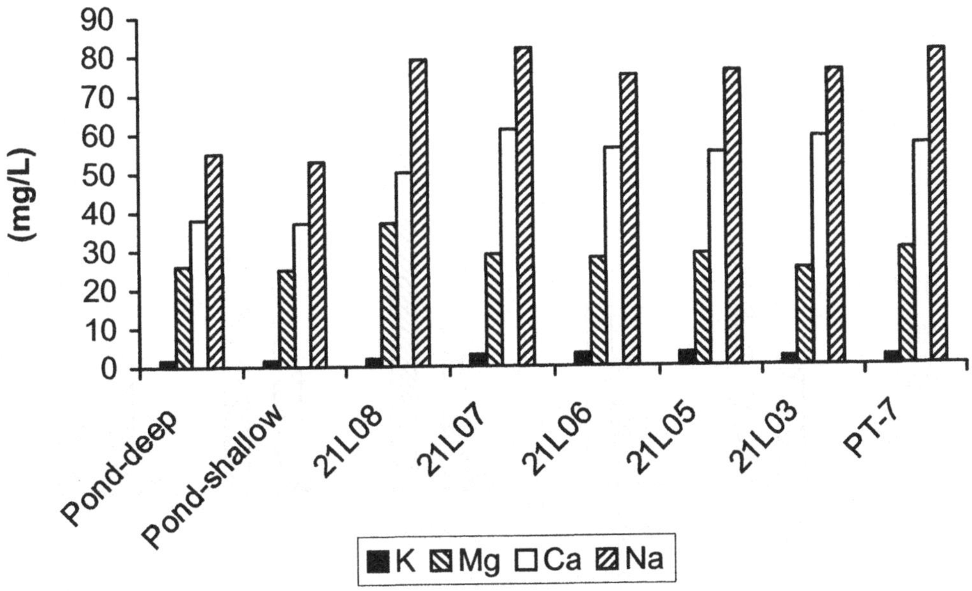

	Kaiser AHF deep	Kaiser AHF shallow	21L08	21L07	21L06	21L05	21L03	PT-7
Well - pond dist. (ft)	0	0	40	80	190	210	690	1190
Ref. App. Table	B-2	B-2	B-2	B-2	B-2	B-2	B-2	B-2
Sampling date	12/29/97	12/29/97	1/13/98	1/13/98	2/4/98	2/4/98	3/3/98	4/7/98
Depth (ft)	62	3						
K (mg/L)	1.7	1.6	2	3.1	3.2	3.3	2.2	2.5
Mg (mg/L)	26	25	37	29	28	29	25	30
Ca (mg/L)	38	37	50	61	56	55	59	57
Na (mg/L)	55	53	79	82	75	76	76	81

Figure 4.21 Major cations in Kaiser Pond and selected down-gradient wells

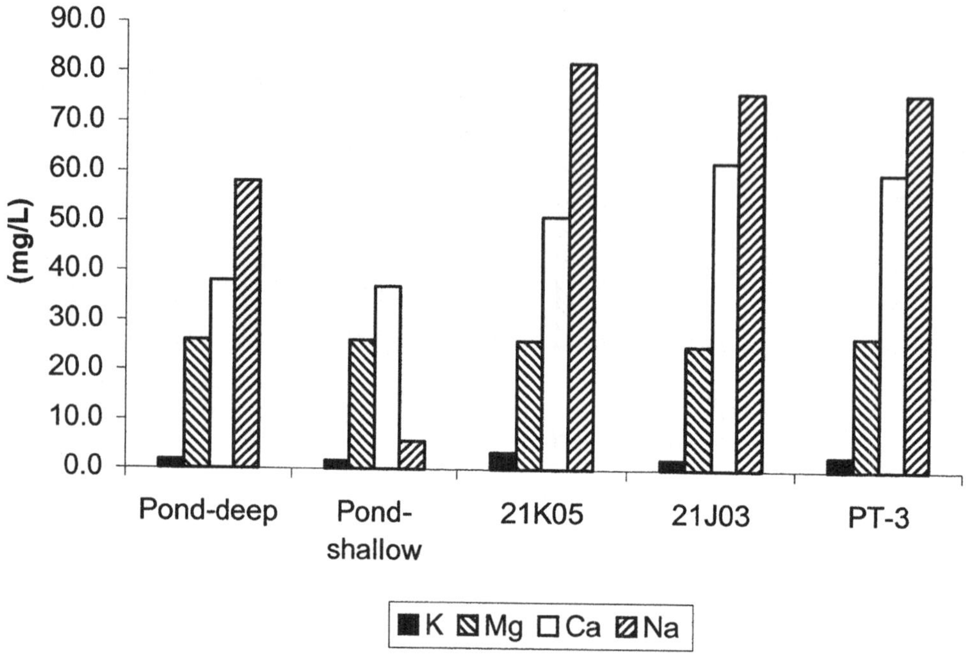

	Bunting Pond-deep	Bunting Pond-shallow	21K05	21J03	PT-3
Pond-well dist. (ft)	0	0	30	50	1800
Ref. App. Table	B-2	B-2	B-2	B-2	B-2
Sample depth (ft)	33	3			
Date collected	12/29/97	12/29/97	2/4/98	3/3/98	6/9/98
K (mg/L)	1.8	1.7	3.4	2	2.8
Mg (mg/L)	26	26	26	25	27
Ca (mg/L)	38	37	51	62	60
Na (mg/L)	58	5.7	82	76	76

Figure 4.22 Major cations in Bunting Pond and selected down-gradient wells

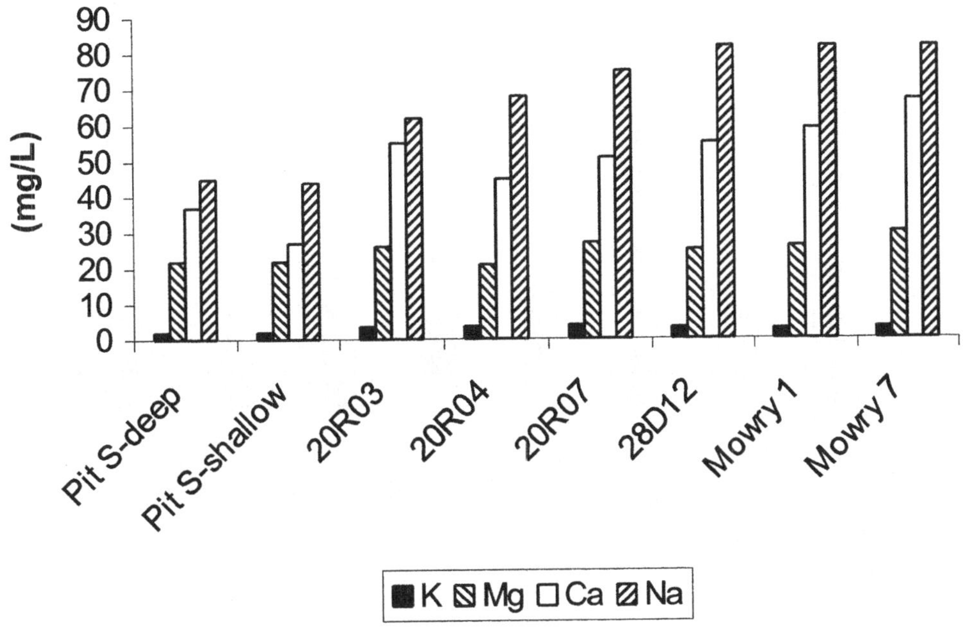

	Pit S deep	Pit S shallow	20R03	20R04	20R07	28D12	Mowry 1	Mowry 7
Closest surface water -Distance (ft)	0	0	Alameda Creek 100	Alameda Creek 100	Alameda Creek 910	Alameda Creek 1740	Kaiser BHF 2740	Kaiser BHF 2750
Ref. App. Table	B-2	B-2	B-2	B-2	B-2	B-2	B-2	B-2
Date	8/25/98	8/25/98	9/25/98	11/17/98	11/17/98	2/24/99	2/24/99	2/24/99
Depth (ft)	40	10						
K (mg/L)	2	1.9	3.4	3.6	3.9	3.2	2.8	3.3
Mg (mg/L)	22	22	26	21	27	25	26	30
Ca (mg/L)	37	27	55	45	51	55	59	67
Na (mg/L)	45	44	62	68	75	82	82	82

Figure 4.23 Major cations in Pit S and BHF wells

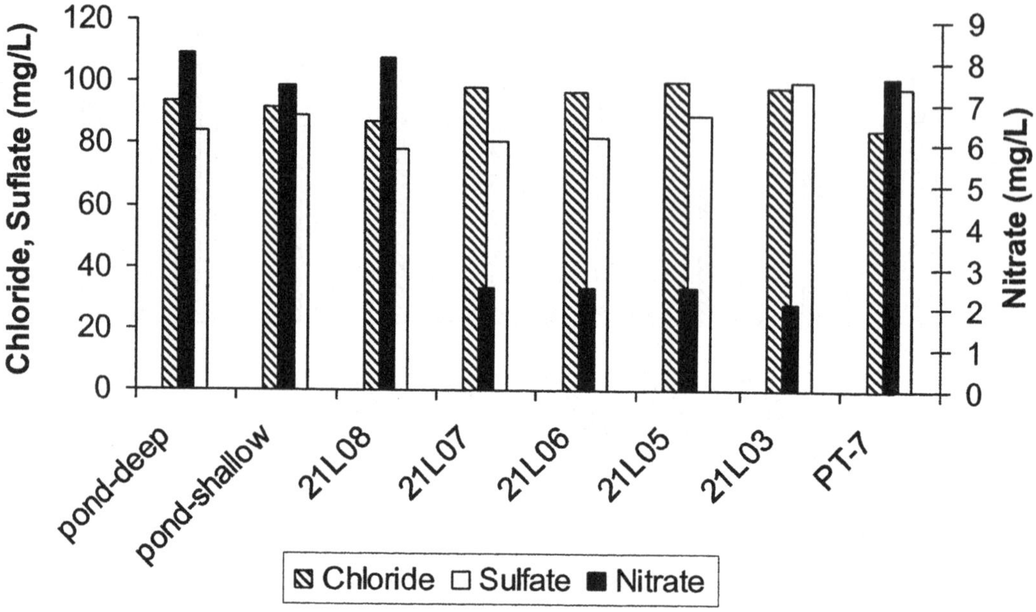

	Kaiser AHF deep	Kaiser AHF shallow	21L08	21L07	21L06	21L05	21L03	PT-7
Well - pond dist. (ft)	0	0	40	80	190	210	690	1190
Ref. App. Table	B-2	B-2	B-2	B-2	B-2	B-2	B-2	B-2
Samping Date	12/29/97	12/29/97	1/13/98	1/13/98	2/4/98	2/4/98	3/3/98	4/7/98
Depth (ft)	62	3						
Nitrate (mg/L)	8.2	7.4	8.1	<5	<5	<5	2.1	7.6
Sulfate (mg/L)	84	89	78	81	82	89	100	98
Chloride (mg/L)	94	92	87	98	97	100	98	85

Figure 4.24 Major anions in Kaiser Pond and selected down-gradient wells

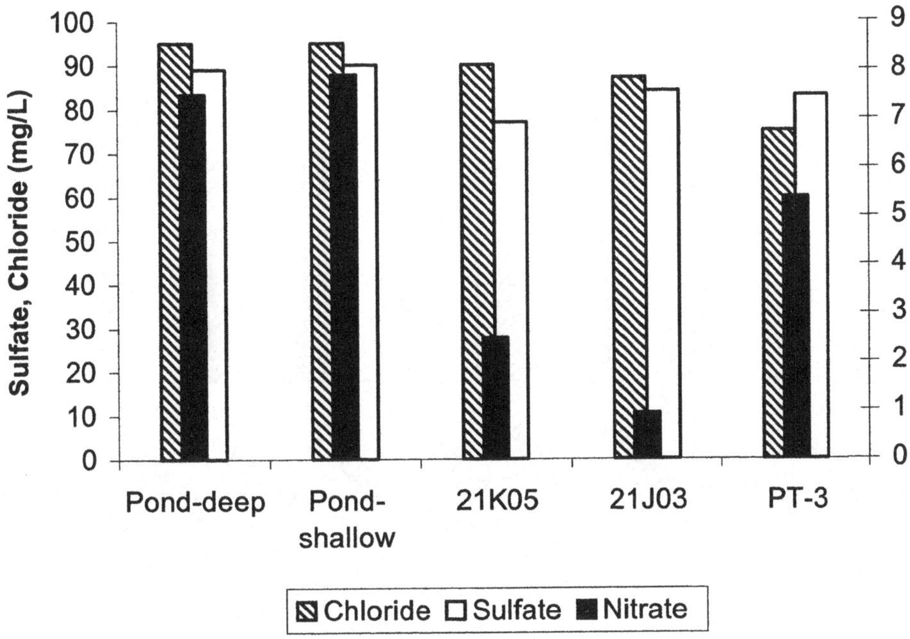

	Bunting Pond-deep	Bunting Pond-shallow	21K05	21J03	PT-3
Pond-well dist. (ft)	0	0	30	50	1800
Sample depth (ft)	33	3			
Ref. App. Table	B-2	B-2	B-2	B-2	B-2
Date collected	12/29/97	12/29/97	2/4/98	3/3/98	6/9/98
Nitrate (mg/L)	7.5	7.9	<5	0.94	5.4
Sulfate (mg/L)	89	90	77	84	83
Chloride (mg/L)	95	95	90	87	75

Figure 4.25 Major anions in Bunting Pond and selected down-gradient wells

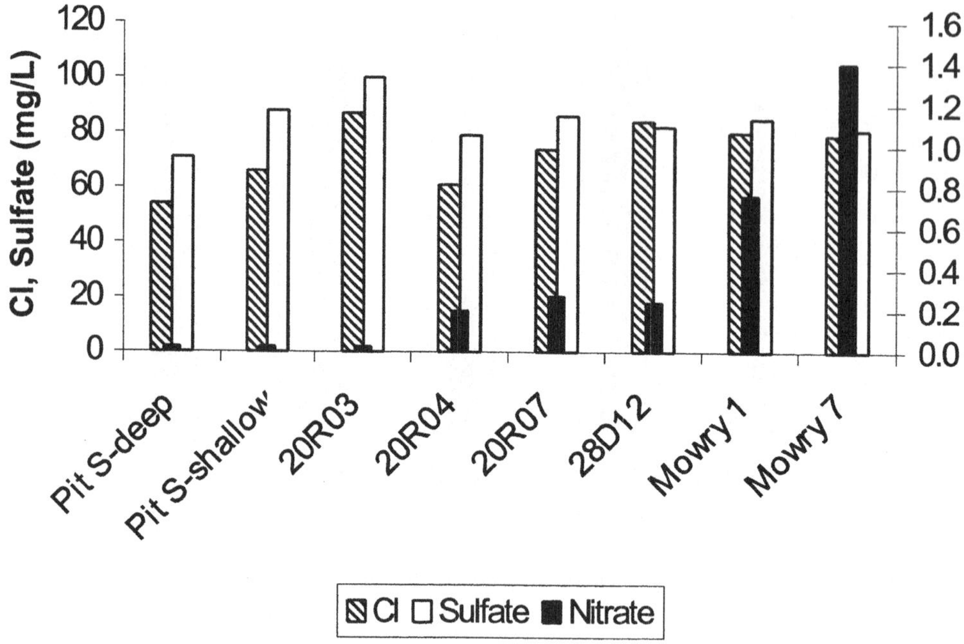

	Pit S deep	Pit S shallow	20R03	20R04	20R07	28D12	Mowry 1	Mowry 7
Closest surface water			Alameda Creek	Alameda Creek	Alameda Creek	Alameda Creek	Kaiser BHF	Kaiser BHF
-Distance (ft)	0	0	100	100	910	1740	2740	2750
Ref. App. Table	B-2	B-2	B-2	B-2	B-2	B-2	B-2	B-2
Date	8/25/98	8/25/98	9/25/98	11/17/98	11/17/98	2/24/99	2/24/99	2/24/99
Depth (ft)	40	10						
Nitrate (mg/L)	<0.05	<0.05	<0.05	0.2	0.27	0.24	0.76	1.4
Sulfate (mg/L)	71	88	100	79	86	82	85	81
Cl (mg/L)	54	66	87	61	74	84	80	79

Figure 4.26 Major anions in Pit S and selected BHF wells

apparent imbalance. As indicated in Table 4.1, dissolved inorganic carbon (DIC) increased from surface water to groundwater. Given the pH range measured in surface water and groundwater samples, bicarbonate should be the dominant species of DIC.

Nitrate

To facilitate a discussion on nitrate, nitrate concentrations in Figures 4.24, 4.25 and 4.26 are repeated in Table 4.12. Dissolved oxygen (DO) data is also included in the tables because the potential for denitrification may be related to diminished levels of DO.

As indicated in Table 4.12, nitrate in Bunting and Kaiser Ponds (AHF) ranged from 7.5 to 8.2 mg/L in December 1997, whereas the levels were below detection (<0.5 mg/L) in Pit S (BHF) in August 1998. The fluctuation could be a consequence of a seasonal limnological process in the recharge ponds, involving algal production (nitrate utilization) in the spring and summer, and cell death and ammonification (nitrate generation) during the fall and winter. Alternatively, the low concentrations in Pit S in August of 1998 may simply be a result of low concentrations in water diverted into the pond from Alameda Creek. During the late summer, a large component of water in Alameda Creek is quarry groundwater extractions (occurring in the Alameda Creek watershed upstream of ACWD) in which nitrate levels may be very low. Nitrate concentrations in Alameda Creek, measured upstream of Rubber Dam No. 3 in August of 1998, ranged from <0.05 to 0.06 mg/L (Table B-2 in Appendix B). The ponds, in general, could be subject to higher nitrogen levels than the creek, and certain individual ponds could be subject to higher nitrate levels than others. Local ecological factors affecting nitrogen budgets in any given pond in the study area may include storm water inputs from pipes or overland drainage, current and historical land use around the pond, and limnological processes.

Groundwater data generally indicate nitrate removal in AHF pond sediments and/or the surrounding aquifer. Although levels in Kaiser and Bunting Pond were not high (under 10 mg/L), the levels in groundwater were generally lower nevertheless. The exception was the 8-mg/L detection at well 21L08. Interestingly, 21L08 had a higher level of dissolved oxygen (4 mg/L) than other wells and the AHF monitoring wells collectively fit a pattern of low DO corresponding with low nitrate levels. This suggests denitrification (an anaerobic process) as a possible important nitrate-removing mechanism. One data point not conforming to this

Table 4.12

Nitrate and dissolved oxygen in surface water and groundwater. From top to bottom: Kaiser Pond and selected down-gradient wells (AHF), Bunting Pond and down-gradient wells (AHF), and Pit S and selected BHF wells

	Kaiser AHF deep	Kaiser AHF shallow	21L08	21L07	21L06	21L05	21L03	PT-7
Well - pond dist. (ft)	0	0	40	80	190	210	690	1190
Ref. App. Table	B-2	B-2	B-2	B-2	B-2	B-2	B-2	B-2
Samping Date	12/29/97	12/29/97	1/13/98	1/13/98	2/4/98	2/4/98	3/3/98	4/7/98
Depth (ft)	62	3						
Nitrate (mg/L)	8.2	7.4	8.1	<5	<5	<5	2.1	7.6
Ref. App. Table	B-1	B-1	B-1	B-1	B-1	B-1	B-1	B-1
Samping date	12/30/97	12/30/97	1/13/98	1/13/98	2/4/98	2/4/98	3/3/98	3/31/98
Depth	54	3						
D.O. (mg/L)	9.1	9.3	4.1	0.6	1.4	0.5	0.6	3.0

	Bunting Pond-deep	Bunting Pond-shallow	21K05	21J03	PT-3
Pond-well dist. (ft)	0	0	30	50	1800
Sample depth (ft)	33	3			
From Appendix Table:	B-2	B-2	B-2	B-2	B-2
Date collected	12/29/97	12/29/97	2/4/98	3/3/98	6/9/98
Nitrate	7.5	7.9	<5	0.94	5.4
From Appendix Table:	B-1	B-1	B-1	B-1	B-1
Date collected	12/30/97	12/30/97	2/4/98	3/3/98	6/9/98
D.O. (mg/L)	10.7	11.1	1.4	1.9	1.2

	Pit S deep	Pit S shallow	20R03	20R04	20R07	28D12	Mowry 1	Mowry 7
Closest surface water -Distance (ft)	0	0	Alameda Creek 100	Alameda Creek 100	Alameda Creek 910	Alameda Creek 1740	Kaiser BHF 2740	Kaiser BHF 2750
Ref. App. Table	B-2	B-2	B-2	B-2	B-2	B-2	B-2	B-2
Date	8/25/98	8/25/98	9/25/98	11/17/98	11/17/98	2/24/99	2/24/99	2/24/99
Depth	40	10						
Nitrate	<0.05	<0.05	<0.05	0.2	0.27	0.24	0.76	1.4
Ref. App. Table	B-1	B-1	B-1	B-1	B-1	B-1	B-1	B-1
Date	8/20/98	8/20/98	9/15/98	11/17/98	11/17/98	2/24/99	2/24/99	2/24/98
Depth	40	3						
D.O. (mg/L)	0.2	7.6	1	0.4	1.3	0.1	1.2	2.1

nitrate-DO pattern for the monitoring wells was the detection of 0.6 and 6.7 mg/L of DO and nitrate, respectively, in well 21L07 on 2/4/98 (see Tables B-1 and B-2 in the Appendix).

The presence of nitrate in PT (AHF) production wells at levels higher than in the monitoring wells suggest the potential for a local source to the production wells other than the recharge ponds. The observation of higher nitrate concentrations may be a consequence of previous agricultural activities or ongoing fertilizing of lawns at nearby residences. Also, because service buildings within the wellfield are served by a sewer line that connects to a main on Mowry Avenue, it could be argued that leakage of sewage may have occurred and increased nitrate levels. However, the non-detects of coliphage during this study, and consistent non-detects of bacteria over the course of regular District monitoring, indicate that either sewage has not leaked or that there is no health risk from the presence of the sewer lines. In addition, the nitrate levels are well below the regulatory drinking water maximum contaminant level of 45 mg/L. The nitrate observed in PT wells may have a natural, subsurface sedimentary source (sedimentary sources of nitrate have been documented in Williams et al. 1998 and Holloway et al. 2001).

With a maximum concentration of 1.4 mg/L, nitrate levels at Mowry (BHF) production wells were even lower than the levels seen at PT (AHF) production wells. However, only two Mowry wells (the only two active wells perforated in the upper aquifer at the time of this study) were sampled for comparison with surface water and monitoring well water. The results do not indicate a significant source of nitrate near the Mowry Wellfield. BHF monitoring wells also had relatively low amounts of nitrate, the maximum level being 0.31 mg/L. Because concentrations in Pit S were also low, there was not sufficient contrast between surface water and groundwater in the BHF sub-basin during the sampling period to comment on the potential capacity for nitrate removal potential for the BHF side of the study area.

Total Hardness, TDS, and Conductivity

Contradicting the cation trend, hardness, TDS, and conductivity did not increase in the transition from surface water to groundwater in the AHF sub-basin (Figures 4.27 and 4.28). However, in the BHF sub-basin, these three parameters were higher in groundwater than in surface water (Figure 4.29), and thus, were consistent with the cation trend.

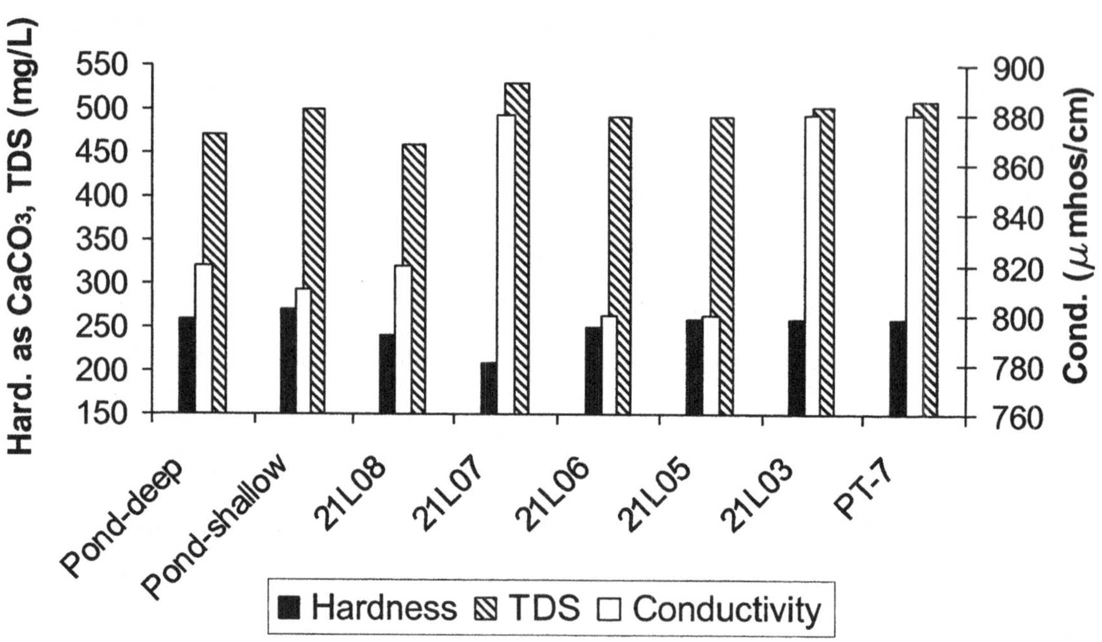

	Kaiser AHF deep	Kaiser AHF shallow	21L08	21L07	21L06	21L05	21L03	PT-7
Well - pond dist. (ft)	0	0	40	80	190	210	690	1190
Ref. App. Table	B-2	B-2	B-2	B-2	B-2	B-2	B-2	B-2
Samping Date	12/29/97	12/29/97	1/13/98	1/13/98	2/4/98	2/4/98	3/3/98	4/7/98
Depth (ft)	62	3						
Hard. (mg/L as $CaCO_3$)	260	270	240	210	250	260	260	260
TDS (mg/L)	470	500	460	530	490	490	502	510
Cond. (μmhos/cm)	820	810	820	880	800	800	880	880

Figure 4.27 Hardness, TDS, and Conductivity in Kaiser Pond and selected down-gradient wells

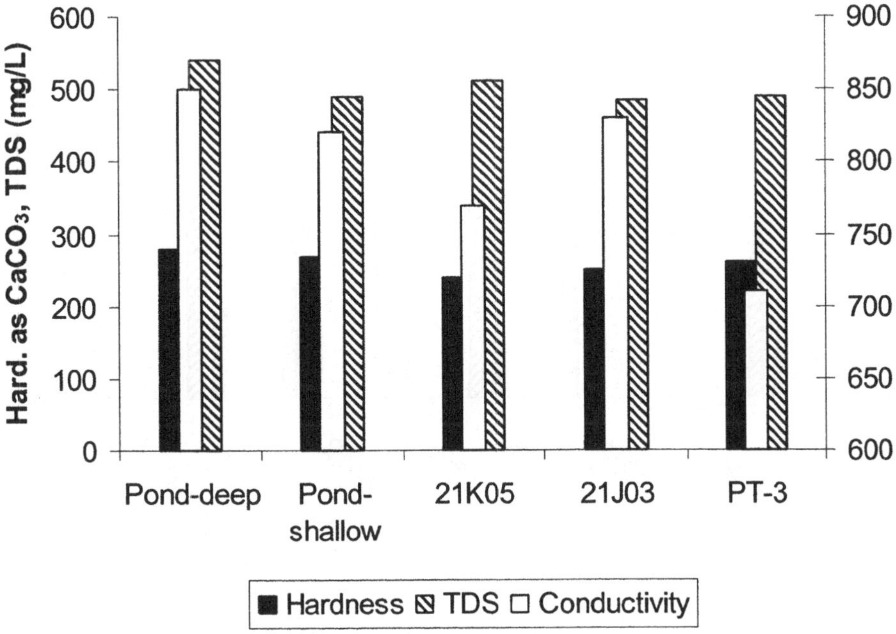

	Bunting Pond-deep	Bunting Pond-shallow	21K05	21J03	PT-3
Pond-well dist. (ft)	0	0	30	50	1800
Ref. App. Table	B-2	B-2	B-2	B-2	B-2
Sample depth (ft)	33	3			
Date collected	12/29/97	12/29/97	2/4/98	3/3/98	6/9/98
Hard. (mg/L as $CaCO_3$)	280	270	240	250	260
TDS (mg/L)	540	490	510	485	490
Cond. (μmhos/cm)	850	820	770	830	710

Figure 4.28 Hardness, TDS, and Conductivity in Bunting Pond and selected down-gradient wells

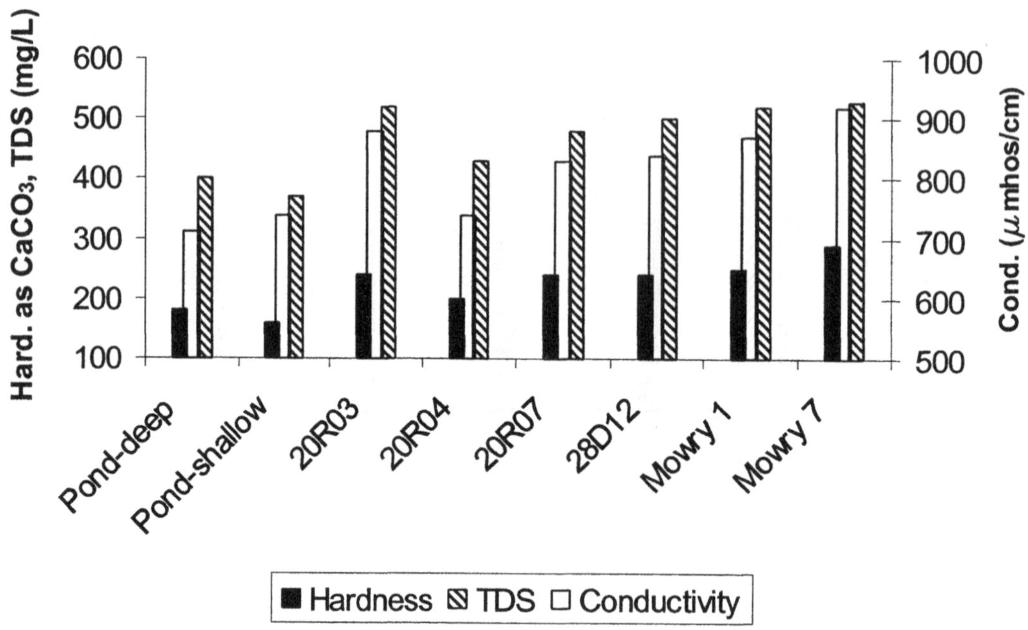

	Pit S deep	Pit S shallow	20R03	20R04	20R07	28D12	Mowry 1	Mowry 7
Closest surface water -Distance (ft)	0	0	Alameda Creek 100	Alameda Creek 100	Alameda Creek 910	Alameda Creek 1740	Kaiser BHF 2740	Kaiser BHF 2750
Ref. App. Table	B-2	B-2	B-2	B-2	B-2	B-2	B-2	B-2
Date	8/25/98	8/25/98	9/25/98	11/17/98	11/17/98	2/24/99	2/24/99	2/24/99
Depth (ft)	40	10						
Hard. (mg/L as $CaCO_3$)	180	160	240	200	240	240	250	290
TDS (mg/L)	400	370	520	430	480	500	520	530
Cond. (μmhos/cm)	710	740	880	740	830	840	870	920

Figure 4.29 Hardness, TDS, and Conductivity in Pit S and selected BHF wells

One explanation for this discrepancy in trends for AHF water is that TDS, hardness, and conductivity may have been measured with less accuracy and/or precision compared to measurements for the individual cations; however, there isn't sufficient information to confirm this. Quality assurance and control reports from the laboratory (note: these analyses were not performed at Lawrence Livermore National Laboratory) were incomplete. All necessary QA/QC protocols were likely followed; however, attempts to recover the necessary additional QA/QC information from the laboratory were unsuccessful because of the amount of time that had elapsed since the analyses. TDS and conductivity were the only parameters consistently included in the QA/QC reports, but cations were not included in any of them. The accuracy and precision errors for detection of TDS in interferent-free matrix solutions, spiked to 500 mg/L, were 8 mg/L and 33 mg/L, respectively. For detection of conductivity in spiked solutions of 710 μmohs/L, accuracy and precision errors were 5 μmhos/cm and 18 μmohs/cm, respectively. Interestingly, the analytical result for the TDS spike associated with the batch of samples from Bunting and Kaiser Ponds was 540 mg/L, 40 mg/L over the known TDS concentration. Hence, TDS for surface water samples could have been over-predicted to an extent that could account for some of the discrepancy between TDS and cation trends. On the other hand, the QA/QC result for conductivity did not exceed the spike concentration for the same batch.

To further investigate the discrepancy, ion balances for AHF samples were checked by converting the major-ion mass concentrations to charge-equivalent concentrations. The bicarbonate ion, though not directly analyzed by the laboratory, was estimated from pH and DIC, or alkalinity and pH. The numerical result was a positive-charge deficiency for all the AHF samples with the exception of well 21L08. The deficiency was greatest for Kaiser and Bunting Pond samples, suggesting that the true concentration of cations in AHF pond water could have been greater than the laboratory result. But the data (see Appendix Table B-2) shows similarity in major cation concentrations among all the surface water samples (AHF and BHF samples combined), including samples collected from Pit S (BHF) in April 1997 and August 1998, from Kaiser and Bunting Ponds (AHF) in December 1997, and from Alameda Creek in August of 1998 (AHF). Among all the surface water samples, calcium was no higher than 41 mg/L, magnesium was less than or equal to 35 mg/L, and sodium did not exceed 67 mg/L. The levels of these cations were generally lower than what was observed in groundwater. The good agreement in cation concentrations of August of 1998 with those measured prior to that time is noteworthy because a different laboratory analyzed the August 1998 samples. This consistency lends support to the argument that dissolution of sedimentary material was occurring; however, further analyses with complete QA/QC reports would be needed to strengthen this argument and to explain the discrepancy between parameter trends.

Alkakinity, DIC, and pH

Alkalinity and pH data for surface and selected groundwater samples are provided in Figures 4.30, 4.31, and 4.32. DIC data are not included in the bar graphs because only a few wells were analyzed for DIC. However, available DIC results are included in the data tables accompanying the bar graphs.

Similar to hardness, TDS, and conductivity, alkalinity did not appreciably increase in the transition from surface water to groundwater in the AHF sub-basin. Moreover, decreases in pH were observed. These observations suggest an aerobic decomposition reaction, which consumes alkalinity, occurring in the pond sediments or the vadose zone, counterbalancing alkalinity-generating processes such as calcium carbonate dissolution and denitrification. Dissolution of soil gas-derived CO_2 into groundwater may also have been occurring. The expected net effect of the combination of these processes is consistent with the observed trends in chemical parameters: an increase in calcium, an increase in DIC, a decrease in pH, and no appreciable change in alkalinity.

The progression of decreasing $\delta^{13}C$ values from the ponds to the production wells (Table 4.1) also supports the notion that dissolution of soil gas-derived CO_2 or in-sediment microbial reactions (such as denitrification and aerobic decomposition of organic matter) are likely dominant sources of the extra DIC in groundwater compared to surface water in this study area. (Refer to Chapter 3 for background on interpreting geo-chemical processes from carbon isotopes.) The notable deficits in dissolved oxygen in groundwater relative to surface water (DO data was presented in Table 4.12) are strongly indicative of an alkalinity-consuming aerobic decomposition reaction in the pond sediments.

Trace Metals

Concentrations of metals were well below health-based regulatory limits and were generally typical for local surface water and groundwater (Tables B-4 and B-6 in Appendix B).

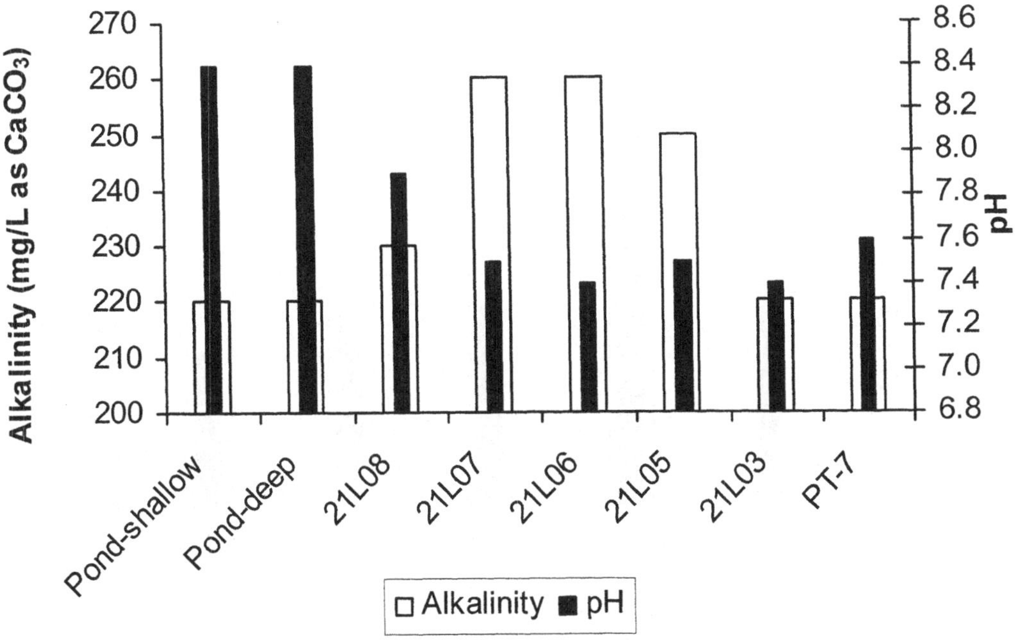

	Kaiser AHF deep	Kaiser AHF shallow	21L08	21L07	21L06	21L05	21L03	PT-7
Well - pond dist. (ft)	0	0	40	80	190	210	690	1190
Ref. App. Table	B-2	B-2	B-2	B-2	B-2	B-2	B-2	B-2
Samping Date	12/29/97	12/29/97	1/13/98	1/13/98	2/4/98	2/4/98	3/3/98	4/7/98
Depth (ft)	3	62						
Alk. (mg/L $CaCO_3$)	220	220	230	260	260	250	220	220
Ref. Table		4.1		4.1	4.1	4.1		4.1
Date		12/29/97		12/16/97	2/4/98	12/2/97		2/4/98
Depth (ft)		24						
DIC (mg/L)	N.A.	252	N.A.	260	296	292	N.A.	299
Ref. App. Table	B-2	B-2	B-1	B-1	B-1	B-1	B-1	B-1
Sampling Date	12/29/97	12/29/97	1/13/98	1/13/98	2/4/98	2/4/98	3/3/98	4/7/98
Depth (ft)	3	62						
pH	8.4	8.4	7.9	7.5	7.4	7.5	7.4	7.6

Figure 4.30 Alkalinity and pH in Kaiser Pond and selected down-gradient wells. Available DIC data provided in data table.

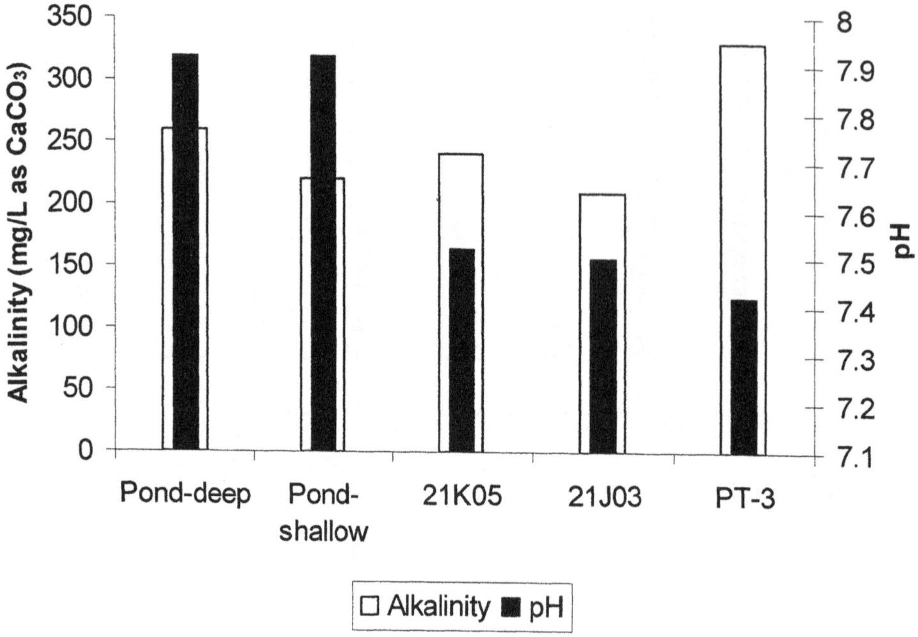

	Bunting Pond-deep	Bunting Pond-shallow	21K05	21J03	PT-3
Pond-well dist. (ft)	0	0	30	50	1800
Ref App. Table	B-2	B-2	B-2	B-2	B-2
Date Collected	12/29/97	12/29/97	2/4/98	3/3/98	6/9/98
Depth	33	3			
Alk. (mg/L as $CaCO_3$)	260	220	240	210	330
Ref. Table		4.1	4.1		
Date Collected		12/30/97	2/4/98		
Depth (ft)		3			
DIC (mg/L)	NA	245	280	NA	NA
Ref App. Table	B-1	B-1	B-2	B-2	B-2
Date Collected	12/30/97	12/30/97	2/4/98	3/3/98	6/9/98
Depth (ft)	33	3			
pH	7.92	7.92	7.52	7.50	7.42

Figure 4.31 Alkalinity and pH in Bunting Pond and selected down-gradient wells. Available DIC data provided in data table.

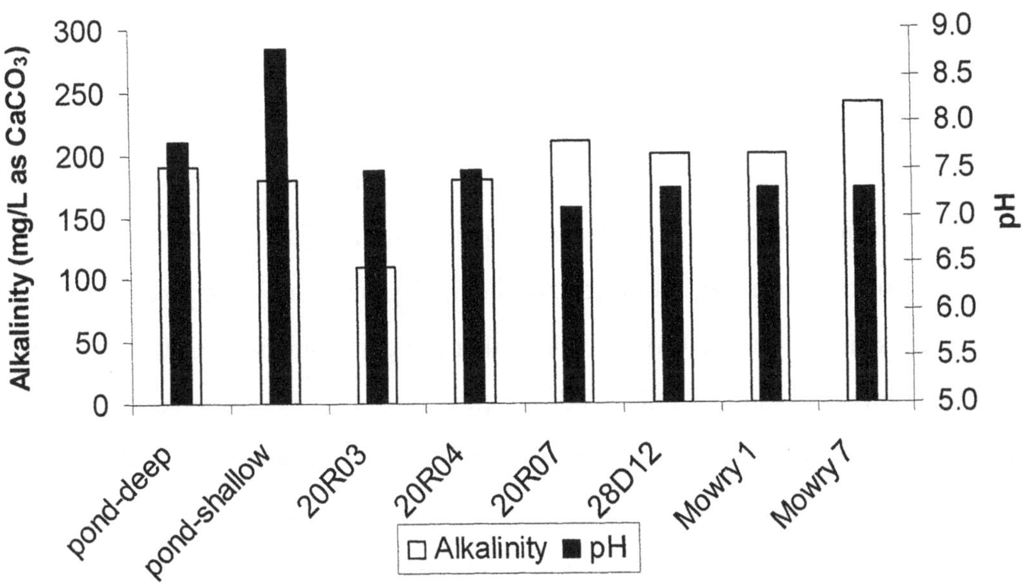

	Pit S deep	Pit S shallow	20R03	20R04	20R07	28D12	Mowry 1	Mowry 7
Closest surface water -distance (ft)	0	0	Alameda Creek 100	Alameda Creek 100	Alameda Creek 910	Alameda Creek 1740	Kaiser BHF 2740	Kaiser BHF 2750
Ref. Table	B-2	B-2	B-2	B-2	B-2	B-2	B-2	B-2
Date	8/25/98	8/25/98	9/25/98	11/17/98	11/17/98	2/24/98	2/24/98	2/24/98
Depth (ft)	40	5						
Alk. as $CaCO_3$ (mg/L)	190	180	110	180	210	200	200	240
Ref. App. Table	B-1	B-1	B-1	B-1	B-1	B-1	B-1	B-1
Date	8/20/98	8/20/98	9/15/98	11/17/98	11/17/98	2/24/99	2/24/99	2/24/98
Depth (ft)	40	3						
pH	7.8	8.8	7.5	7.5	7.1	7.3	7.3	7.3
Ref. Table							4.1	4.1
Date							6/30/97	4/16/97
DIC (mg/L)	N.A.	N.A.	N.A.	N.A.	N.A.	N.A.	354	370

Figure 4.32 Alkalinity and pH in Pit S and selected BHF wells. Available DIC data provided in data table.

Several trace metals appear to have behaved conservatively during infiltration, including Rb, Sr, Mo, Ag, Cd, Ba, Pb, Cr and Co. This suggests that dissolution of sedimentary material during recharge is not a significant source of these metals in this study area, nor do the sediments provide a sink for trace metals through sorption during infiltration. The lack of increase in Sr and Ba is in contrast with the observed increase in Ca, as described above, and suggests that the carbonate mineral interacting with infiltrating water is not aragonite or dolomite, but rather a pure form of calcite. Pb and Cd are particle reactive metals, and a concentration change is expected with a significant change in particulate concentration. The observed lack of change in the concentrations of these metals (measured on the 0.45 μm filtrate) suggests that the particle-bound component is small. Furthermore, the fact that the concentrations of the trace metals can be explained by considerations of ORP and pH (see next two paragraphs) suggests that the metals were present in dissolved forms, and that colloidal fractions do not play a large role. Jensen et al. (1998) performed ultrafiltration on groundwater samples from reducing environments, and found that several trace metals, including Mn and Fe, were present almost exclusively in the truly dissolved (<.001 μm) form.

The changes in trace metal concentrations that take place during recharge can largely be attributed to increased solubility associated with a decrease in pH and/or oxidation-reduction potential. As indicated in Figures 4.30, 4.31, and 4.32, a significant drop in pH was observed in the transition from surface water to groundwater in this study, and was attributed largely to microbial oxidation of organic carbon. Several studies have documented that a bacteria-mediated oxidative decomposition of organic matter in sediments leads to these chemical changes in infiltrating groundwater.

Concentrations of aluminum, iron, and manganese in some samples were one or more orders of magnitude higher than in others (for example, in Table B-4, compare the aluminum concentration in well 21L07 on 12/2/97 and 2/4/98). One reason for this may be that the fraction measured was not strictly the dissolved component: the samples were acidified in the field and filtered upon return to the laboratory. Therefore, some particle-bound cations may have been released into solution. This procedure was carried out to be consistent with typical monitoring procedures for drinking water. However, samples collected from Pit S and wells 20R03 and 20J04 (a subset from the BHF portion of the study) were filtered upon return to the laboratory and then acidified. Al and Mn concentrations were indeed significantly lower in these latter four samples, while Fe concentrations were in the same range as the rest of the data set.

Measurements of oxidation-reduction potential and iron and manganese concentrations in Pit S and wells 20R03 and 20J04 are presented in the bar graph and accompanying data table of Figure 4.33. The hypolimnion of Pit S had the lowest DO and ORP, and as would be expected, and the highest levels of manganese and iron. More moderate levels of Fe (less than 200 ug/L) and Mn (0.35 to to 3.6 ug/L) in shallow pond water and groundwater corresponded with ORP values between 239 and 269 mV. Generally, field-measured ORP correlated most closely with iron, somewhat closely with Mn, but not very well with dissolved oxygen.

Measured ORP values were compared with theoretical ORP values calculated from pH, Mn, Fe, and DO. The following half reactions were considered.

$\frac{1}{4} O_2 + H^+ + e^- = \frac{1}{2} H_2O$,	$p\varepsilon^0 = 20.75$	(4.1)
$\frac{1}{2} MnO_2\ (s) + 2H^+ + e^- = \frac{1}{2} Mn^{+2} + H_2O$	$p\varepsilon^0 = 20.80$	(4.2)
$Fe(OH)_3\ (am) + 3H^+ + e^- = Fe^{+2} + 3H_2O$	$p\varepsilon^0 = 16.2$	(4.3)

The ranges of calculated oxidation-reduction potentials are 690 to 760 mV (based on Equation 4.1 and measured DO and pH), 400 to 580 mV (based on Equation 4.2 and measured Mn and pH), and -280 to -40 mV (based on Equation 4.3 and measured Fe and pH). In order to be controlled by field-measured ORP, dissolved oxygen and iron would have to be less (iron would not necessarily need to be less depending on how much of it was complexed instead of existing in the form of free Fe^{+2}) and manganese greater than the measured values in all four samples. However, given the probable amount of carbonate ion in the samples, a lack of $MnCO_3(s)$ solubility would prevent Mn from reaching levels in equilibrium with $MnO_2(s)$ as estimated by Equation 4.2 (with input of field-measured ORP and pH). Concentrations of carbonate ion were estimated by the alkalinity and pH, assuming that pH in the system was buffered primarily by the carbonate system. Based on these estimates, the maximum Mn^{+2} concentration in equilibrium with $MnCO_3(s)$ would be 2,500 ug/L, 280 ug/L, 8,500 ug/L, and 4,500 ug/L in the deep sample from Pit S, the shallow sample from Pit S, and samples from 20R03 and 20R04, respectively. If field-measured ORP values are close to the actual ORP that controlled levels of dissolved manganese in these samples, it could be concluded that manganese

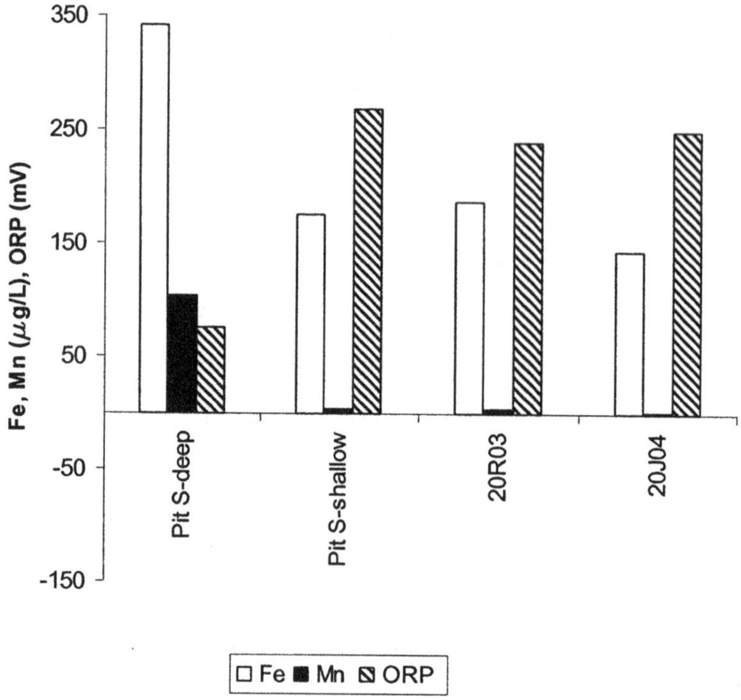

	Pit S deep	Pit S shallow	20R03	20J04
Closest surface water -Distance (ft)	0	0	Alameda Creek 100	Alameda Creek 100
Ref. App. Table:	B-6	B-6	B-6	B-6
Date	8/25/98	8/25/98	9/15/98	9/15/98
Depth (ft)	40	5		
Fe (µg/L)	342	176	187	143
Mn (µg/L)	103	4	3	0.4
Ref. App. Table:	B-1	B-1	B-1	B-1
Date	8/20/98	8/20/98	9/15/98	9/15/98
Depth (ft)	40	3		
pH	7.8	8.8	7.5	7.5
ORP (mV)	75	269	239	248
D.O. (mg/L)	0.2	7.6	1.0	0.4

Figure 4.33 Iron, manganese and oxidation-reduction potential in Pit S and wells 20R03 and 20R04. Dissolved oxygen data also provided in data table.

in surface water and groundwater in the BHF portion of the study area was under-saturated with respect to all potential manganese precipitates. On the other hand, there may have been mixed potentials in solution, and manganese could have been saturated with respect to $MnO_2(s)$ if manganese speciation was under the influence of higher potentials (over 400 mV). The field-measured ORP values may have been reflecting a mixed solution potential, and the ORP probe may have been sensitive to a composite effect exerted by iron and manganese or other redox species. The limitations of direct measurements of ORP in natural aquatic systems is well documented (Drever 1988). Unlike species in acid-base and complexation reactions, the concentrations of species that make up redox couples in natural waters are not often well predicted by equilibrium models. ORP reactions are often rate-dependent and/or microbially-mediated, and a solution cannot always be characterized by a single ORP value.

Organic Compounds

Data for MTBE and TOC are presented in Figures 4.34, 4.35, and 4.36. TOC may incorporate mutagenic compounds and increases trihalomethane formation potential; therefore, levels above a few milligrams per liter are undesirable in drinking water. Results of certain AHF samples collected for TOC analysis between 5/8/97 and 3/3/98 appear to be incorrectly high (Table B-2). Re-runs of samples by the contract laboratory resulted in lower values. Split samples analyzed by ACWD's in-house laboratory were also substantially lower than the contract laboratory's initial results. Analyses of samples collected from the BHF portion of the study area in 1998 were performed entirely at LLNL (LLNL did not analyze AHF samples for TOC). TOC data for the BHF side is considered more reliable than the AHF side due, in part, to the care taken at LLNL to completely remove the inorganic carbon in the water before analyzing for TOC. The smaller TOC values in BHF samples, relative to AHF samples, are probably a consequence of the removal of DIC at the laboratory.

Figures 4.34, 4.35, and 4.36 indicate a significant reduction of TOC in the transition from surface water to groundwater. Decreases in TOC during recharge have been documented at artificial recharge facilities in Orange County, Calif. (Davisson et al. 1998) and in Mesa, Ariz. (Drewes and Fox 1999a). Results of this study show similar percentage decreases (~50 percent), with lower starting concentrations. Several possible mechanisms for chemical transformation or physical removal of TOC are postulated (Drewes and Fox 1999b, Hessen et al. 1997, Davisson et

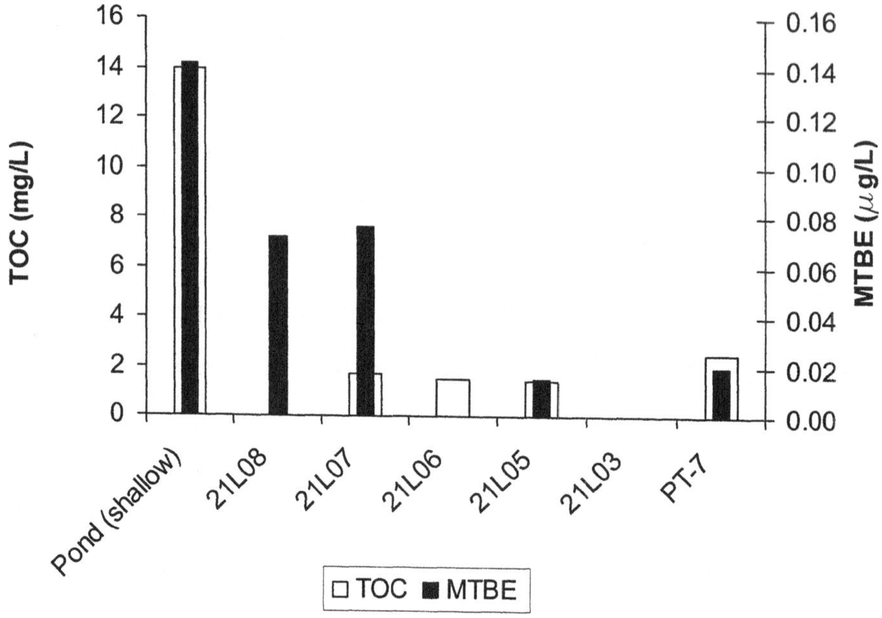

	Kaiser AHF deep	Kaiser AHF shallow	21L08	21L07	21L06	21L05	21L03	PT-7
Well - pond dist. (ft)	0	0	40	80	190	210	690	1190
Date Collected	12/29/97	12/29/97	1/13/98	1/13/98	2/4/98	2/4/98	3/3/98	4/7/98
Ref. App. Table	B-2	B-2	B-2	B-2	B-2	B-2	B-2	B-2
Depth (ft)	36	3						
TOC (mg/L)	6.3	14	NA	1.7	1.5	1.4	NA	2.5
Ref. App. Table	B-3	B-3	B-3	B-3	B-3	B-3	B-3	B-3
Depth (ft)	23	3						
MTBE (µg/L)	0.127	0.142	0.072	0.076	NA	1.5	ND	2

Figure 4.34 Total Organic Carbon and MTBE in Kaiser Pond and selected down-gradient wells

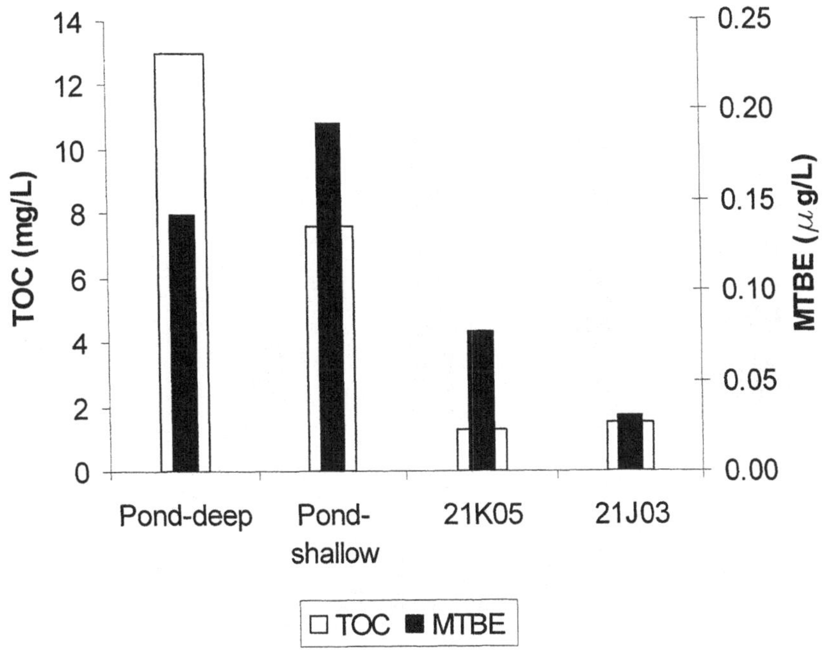

	Bunting Pond-deep	Bunting Pond-shallow	21K05	21J03	PT-3
Pond-well dist. (ft)	0	0	30	50	
Sample depth (ft)	33	3			
Date collected	12/29/97	12/29/97	2/4/98	3/3/98	NA
Ref. App. Table	B-2	B-2	B-2	B-2	NA
TOC (mg/L)	13	7.6	1.3	1.5	NA
Ref. App. Table	B-3	B-3	B-3	B-3	NA
MTBE (μg/L)	0.142	0.193	0.077	0.031	NA

Figure 4.35 Total Organic Carbon and MTBE in Bunting Pond and selected down-gradient wells

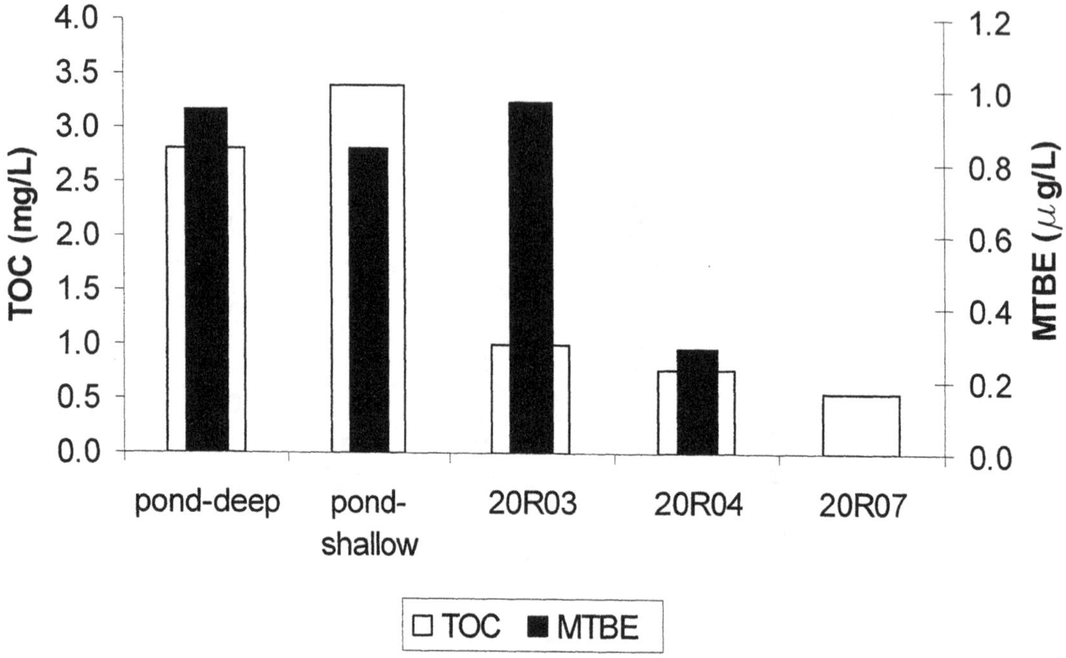

	Pit S pond-deep	Pit S pond-shallow	20R03	20R04	20R07	28D12	Mowry 1	Mowry 7
Closest surface water -Distance (ft)	0	0	Alameda Creek 100	Alameda Creek 100	Alameda Creek 910	Alameda Creek 1740	Kaiser BHF 2740	Kaiser BHF 2750
Ref. App. Table	B-3	B-3	B-3	B-3	B-3			
Date	8/25/98	8/25/98	9/15/98	11/17/98	11/17/98			
Depth (ft)	40	10						
TOC (mg/L)	2.8	3.4	1	0.76	0.56	NA	NA	NA
MTBE (μg/L)	0.95	0.84	0.97	0.29	NA	NA	NA	NA

Figure 4.36 Total Organic Carbon and MTBE in Pit S and selected BHF wells

al. 1998). Dissolved organic carbon may be photo-oxidized by UV-B radiation, or converted to CO_2 via bacterial oxidation (Hessen et al. 1997). With reference to the observed reduction in pH, increase in DIC, as well as the decrease in DO and nitrate, the authors believe that a combination of aerobic oxidative decomposition and denitrification processes was occurring in the sediments, and is at least partially responsible for the observed decrease in TOC.

TOC has several components, and a major fraction is often unidentified compounds. A labile component of the organic material may be re-mineralized during recharge, while a recalcitrant component survives on longer time scales. For example, Davisson et al. (1998) documented a decrease in the ^{14}C content of the TOC during recharge, implying loss of the younger component. Interactions with sedimentary organic matter near the water/sediment interface may also change the ^{14}C content by introducing old, refractory carbon.

Drewes and Fox (1999b) examined transformations in the bulk characteristics of the carbon during recharge using UV absorbance. They found that aromatic compounds are either preferentially removed or converted to non-aromatic products, so that the organic carbon of the distal groundwater has low aromaticity. In areas where treated wastewater is artificially recharged (Drewes et al. 1999), the surviving organic component is found to be the natural organic matter – anthropogenic components are removed or transformed by bacteriological processes in soils. More research is necessary to determine whether the surviving components of TOC are less conducive to the formation of trihalomethanes.

Surface water from four locations in the ACWD watershed was analyzed for several specific trace organic compounds, with the intent of finding an organic tracer for surface water/groundwater interaction. Potential candidates for tracers of groundwater contamination that were found in sub-part per billion concentrations, include simazine (a herbicide), oxadiazon (a herbicide often used on golf courses), and diazinon (an insecticide). These compounds were found at only one or two of the four surface water locations tested, and at concentrations close to the detection limit. Therefore, they were not pursued as potential tracers.

In addition, an analytical method was developed for determination of ultra low-level concentrations of the gasoline additive, methyl *tert*-butyl ether (MTBE). The purge-and-trap, GC-MS (ion selective monitoring mode) technique has a detection limit of 15 parts per trillion. The method is described in detail in Ekwurzel et al. (2001). MTBE was detected at sub-ppb levels in nearly all AHF surface water and groundwater samples. The indications that the source of MTBE is above ground include the very low concentrations at which MTBE was detected, the patterns of decreasing concentrations with depth in surface waters, and a pattern of decreasing concentrations in groundwater away from the ponds.

Concentrations measured in the ponds are similar to those measured in local rainwater (100 ppt), suggesting that precipitation is the likely non-point, surface source. Furthermore, the values measured in precipitation are within the expected range according to the Henry's Gas Law constant, the temperature range during the precipitation events, and published values (~ 1 to 2 ppbv) of atmospheric MTBE concentrations (Brown 1997). However, a contribution of MTBE from motorized watercraft cannot be ruled out. Although there is no (gasoline powered) motorized watercraft use in ACWD's recharge facilities, a significant quantity of ACWD's recharge water is imported from the Sacramento-San Joaquin Delta and the Del Valle Reservoir where use of motorized water craft is seasonally heavy. A subsurface source can be ruled out by comparing the levels and spatial distribution in samples measured from ACWD facilities with those measured near sub-surface point sources. While waters impacted by point sources often have levels in the part-per billion (and higher) range, ACWD waters were all well below 1 ppb.

Further evidence that the MTBE present in ACWD facilities is of atmospheric origin comes from a study of a runoff event in Alameda Creek (Ekwurzel et al. 2001). In that study, pre-event groundwater is distinguished from overland flow in the hydrograph of the creek, and the maximum MTBE concentration (100 ppt) closely matches the concentration in meteoric water from the event. The MTBE concentration in precipitation is likely to vary temporally and spatially according to local air concentration and frequency and intensity of precipitation. The survey of surface water from the San Francisco Bay area, reported in Ekwurzel et al. (2001), suggests that concentrations below 200 ppt reflect atmospheric input only. This range is similar to those reported for atmospheric input by Squillace et al. (1996) for the USGS NAWQA study, and in an USEPA (1998) overview of sources of MTBE. Thus, the values measured in ACWD surface water, all of which were below 200 ppt, can be attributed to atmospheric fallout.

The decrease in MTBE concentrations in monitoring wells located near the surface water stations suggest degradation may be occurring, but some MTBE remains recalcitrant. Several recent studies have concluded that natural biodegradation of MTBE takes place in various settings (Erickson 2001, Schirmer et al. 1999). Alternatively, there is a significant correlation between the MTBE concentration in wells and tracer concentration, suggesting that the decrease in MTBE in groundwater is due to mixing with slightly older water with a lower MTBE concentration. Steep increases in the MTBE content in California gasoline took place in the late

1980's and early 1990's, especially after January 1995 when amendments to the Clean Air Act required metropolitan areas (such as the San Francisco Bay area) to use reformulated gasoline containing fuel oxygenates year-round (Squillace et al. 1996). By the time surface waters are drawn into large production wells that also draw deeper and older water into the well column, the MTBE is diluted down to levels at or below the detection limit. Therefore, the pattern of MTBE occurrence in wells supports the conclusions of the isotope data; i.e., recently recharged water is mixed with older groundwater (having lower MTBE concentrations).

Coliphage

Concentrations of male-specific coliphage (hereinafter referred to as coliphage) in Kaiser and Bunting Ponds (AHF) in December 1997 ranged from non-detect to 3 plaque-forming units (pfu)/100 ml in Bunting Pond and 7 pfu/100 ml in Kaiser Pond. Pit S (BHF) was sampled in August 1998 and coliphage was below detectable levels. In groundwater, phage was not detected in any of the samples collected during this study (Table 4.13).

Many of the wells sampled for coliphage are located within only tens of feet from the ponds. The contrast between surface water and groundwater (some detections versus no-detections) offer some indication that pond sediments or nearby aquifer media may be preventing entry of coliphage into groundwater. Sediment properties thought to be characteristic of ACWD ponds, including high organic content, low porosity, and a transition toward lower pH, are consistent with immobilization or attenuation of virus particles, documented in studies on virus fate and transport (referenced in Chapter 1). On the other hand, the low concentrations of coliphage in shallow AHF pond water (<1 pfu/100ml at the 1-meter depth level in Kaiser Pond and <1 to 3 pfu/100 ml in the 1-meter depth in Bunting Pond), and the complete absence of detectable coliphage in Pit S, preclude any more meaningful conclusions. Dosing the ponds with coliphage would have been required to bring the concentrations up to a workable level to provide a more meaningful estimate of the capacity of pond sediments to remove viruses from pond water percolating to groundwater.

The low concentrations of coliphage in surface water suggests Alameda Creek, including segments in the watershed upstream of ACWD, is not subject to significant sources of fecal contamination. The absence of coliphage in groundwater samples collected for this study is consistent with results from an earlier EPA-sponsored study (in which ACWD was a participant)

Table 4.13

MS Coliphage in surface water and groundwater. From top to bottom: Kaiser Pond and selected down-gradient wells (AHF), Bunting Pond and down-gradient wells (AHF), and Pit S and selected BHF wells

	Kaiser pond-deep	Kaiser pond-shallow	21L08	21L07	21L06	21L05	21L03	PT-7
Well - pond dist. (ft)	0	0	40	80	190	210	690	1190
Ref. App. Table	B-2	B-2	B-2	B-2	B-2	B-2	B-2	B-2
Samping Date	12/29/97	12/29/97	1/13/98	1/13/98	2/4/98	2/4/98	3/3/98	4/7/98
Depth (ft)	62	3						
MS Coliph. (PFU/100 mL)	7	<1	<1	<1	<1	<1	<1	<1

	Bunting Pond-deep	Bunting Pond-shallow	21K05	21J03	PT-3
Pond-well dist. (ft)	0	0	30	50	1800
Sample depth (ft)	33	3			
Ref. App.Table:	B-2	B-2	B-2	B-2	B-2
Date collected	12/29/97	12/29/97	2/4/98	3/3/98	6/9/98
MS Coliph. (PFU/100 mL)	1	<1 to 3	<1	<1	<1

	Pit S deep	Pit S shallow	20R03	20R04	20R07	28D12	Mowry 1	Mowry 7
Closest surface water -Distance (ft)	0	0	Alameda Creek 100	Alameda Creek 100	Alameda Creek 910	Alameda Creek 1740	Kaiser BHF 2740	Kaiser BHF 2750
Ref. App. Table	B-2	B-2	B-2	B-2		B-2	B-2	B-2
Date	8/25/98	8/25/98	9/25/98	11/17/98		2/24/99	2/24/99	2/24/99
Depth	40	10						
MS Coliph. (PFU/100 mL)	<1	<1	<1	<1	N.A.	<1	<1	<1

on the vulnerability of municipal drinking water wells to viral and bacterial contamination (International Consultants 1998). For the EPA study, samples were collected in July 1997 from Mowry 3, Mowry 7, and PT-5, and analyzed for a comprehensive suite of viral and bacteriological parameters, including total coliforms, *Enterococci*, coliphage, Rotavirus, Hepatitus A virus, and Enterovirus. None of the analytes were detected in any of the three ACWD wells.

CHAPTER 5
SUMMARY AND CONCLUSIONS

BACKGROUND AND OBJECTIVES

The purpose of the project was to demonstrate the value of age-dating and isotopic tracers in characterizing the flow dynamics and water quality changes in a complex groundwater domain that includes artificial recharge facilities and pumping wells. Documentation of water quality changes that occur during recharge and transport in groundwater was also an objective of this investigation.

The study area for this project is located at the apex of the Niles Cone, an alluvial fan adjacent to the southeast corner of San Francisco Bay. The Hayward Fault acts as a hydraulic barrier to lateral groundwater flow, effectively cutting the study area into two independent flow domains, the Above-Hayward Fault (AHF) and Below-Hayward Fault sub-basins. Each sub-basin is artificially recharged with surface water that is conveyed by Alameda Creek and then diverted into deep recharge ponds. A major wellfield is located in each sub-basin, with the Mowry and Peralta-Tyson (PT) Wellfields located in the AHF and BHF sub-basins, respectively. The aquifers in both sub-basins are highly permeable, but recharge from the ponds is restricted by a fine-grained layer of sediment on the bottom and sides of the ponds. The AHF sub-basin is considered as having only a single major aquifer layer; however, a thin layer of fine-grained soil is known to exist in the PT Wellfield. This fine-grained layer, though probably not laterally extensive throughout the sub-basin, may act as an aquitard within the AHF portion of the study area. In the BHF sub-basin, four major aquifer layers exist. By order of depth, they are the Newark Aquifer, the Centerville Aquifer, the Fremont Aquifer, and the Deep Aquifer. Within the study area, the Newark Aquifer (BHF) and AHF Aquifer are unconfined. Because of their hydraulic contact with the recharge ponds, these two aquifers were examined the most thoroughly in this study.

METHODS

The initial study portion of the project involved sampling selected groundwater wells and recharge water for natural isotopes. Groundwater ages in various wells were calculated by the tritium-helium method. Oxygen-18 and deuterium were also measured in an attempt to differentiate between local versus imported sources of recharge from which groundwater in various locations of the study area is derived. Carbon-13 and Carbon-14 content was measured to provide some additional insight on age of the water, and to infer geochemical processes that contribute to dissolved inorganic carbon in groundwater. In addition, the dominant isotopes of dissolved atmospheric gases (He, Ne, Ar, Kr, and Xe) were measured in AHF wells in order to estimate the temperature of the water at the time it was recharged. The abundances of atmospheric gases were also measured for the purpose of determining the amount of dissolved air in groundwater.

Tracer experiments followed the initial study. The artificial isotope tracers that were loaded into ponds provided quantitative information on spatial and temporal variability in recharge and on the rate at which groundwater moves from the ponds to wells. In addition, the parameters that affect the rate of flow and the spatial distribution of fast flow paths were inferred from tracer data. Multiple tracers consisting of rare isotopes of noble gases were used to track groundwater flow from Alameda Creek and individual recharge facilities. Tracers used in the AHF sub-basin included ^{136}Xe and ^{124}Xe, added to Kaiser and Bunting ponds, respectively, and ^{129}Xe, added to the AHF segment of Alameda Creek. In the BHF part of the study area, ^{22}Ne was added to Stevenson Pond and the 'Pit S' portion of Lago los Osos. Tracer concentrations in surface water were recorded during the dosing period and for three weeks thereafter. The progress of tracer migration in groundwater was tracked over a period of several months.

Groundwater and surface water samples were collected for analysis of complete water chemistry and coliphage. Samples for this characterization were collected at strategic points in time in the attempt to coincide with the movement of the tracer center of mass. The purpose of this timing was to gain insight on water quality changes in a mass of recharge water flowing from one point in the aquifer to the next. Also of interest were changes in water quality as surface water percolates through the layer of pond sediment and/or the vadose zone. Such changes were inferred through contrasts in water chemistry between pond surface water and groundwater sampled from monitoring wells close to the ponds.

RESULTS AND CONCLUSIONS

Initial Study

Groundwater ages, determined by the tritium-helium age dating method, varied from less than one year to 31 years for monitoring and production wells. Production wells from the Above Hayward Fault sub-basin were determined to have relatively young ages, with a mean of 5.7 years. Below Hayward Fault production wells that are screened in the middle and deep aquifers have considerably older mean ages of about 20 and 30 years, respectively. These ages represent the mean residence time in the basin for water drawn into the wells.

In the AHF and BHF sub-basins, environmental isotopes of Oxygen-18 and deuterium indicate that groundwater in the study area has been derived mostly from the local Alameda Creek watershed. Surface water in certain large BHF recharge ponds was more enriched in these isotopes relative to other ponds, indicating that evaporation in the Quarry Pits complex is fairly significant. The trend of the negative values of $\delta^{13}C$ and ^{14}C levels below 100 pmC suggests mixing of young and older water in AHF wells, increasing age with depth of aquifer layer (BHF), and possible dissolution of carbonate minerals (in both AHF and BHF sub-basins).

Analyses of dissolved atmospheric gasses indicate that air was not significantly entrained in recharge water that percolated through the vadose zone in the AHF Sub-basin. The lack of variation in calculated time-of-recharge temperatures among groundwater samples from various wells suggests that seasonal surface water temperature variations were not preserved in groundwater. This, in turn, would suggest that recharge occurs slowly enough to equilibrate with soil temperatures.

Tracer Experiments

Hydrologic Conditions during Tracer Studies

Interpretation of the observed trends in tracer concentrations in AHF and BHF surface water and groundwater required close examination of hydrologic conditions in the sub-basin and in the ponds. Hydrologic parameters tracked during the tracer studies included groundwater pumping, pond surface water elevations, and groundwater piezometric head. Rates of seepage (recharge) from Kaiser Pond were calculated using a mass balance approach. The attempt to explain seepage was not central to the research objectives of this study, but some site-specific investigation was needed nevertheless to guide the discussion of the fate of tracer after dosing the recharge ponds.

Seepage rates were compared with the groundwater heads, pond levels and inflow rates to determine their influence on seepage. The analysis indicated poor correlation of seepage rates with pond and groundwater heads. There was a strong correlation with rates of surface water inflow to the pond, although the numeric dependency of seepage rates to inflow rates in the mass balance equation may have, to some extent, falsely preordained such a high correlation. A comparison of PT Wellfield pumping rates with AHF groundwater head demonstrated that AHF well levels are very responsive to PT Wellfield pumping, including wells that are very close to the shorelines of the ponds. Because pond levels track closely with PT Wellfield pumping but pond seepage does not correlate well with aquifer head, it can be concluded that the PT Wellfield has little influence on Kaiser Pond recharge. These results indicate that wellfield pumping controls groundwater movement, while percolation is controlled by the clogging layer on the bottom and sides of the pond beneath the water surface. Percolation probably occurs mainly at very shallow depths where the clogging layer may not be as pervasive due to occasional drying and resistance to sedimentation of fine-grained soil. Therefore, although the AHF pond tracers were injected deep in the ponds, below the elevation of peizometric head of surrounding groundwater, most of the tracer mass was likely fluxed to groundwater near the pond shoreline and at very shallow depths. The mechanics of seepage at Bunting Pond is probably similar to those controlling percolation at Kaiser Pond, but findings from the analysis of Kaiser Pond are not necessarily applicable to BHF ponds.

Surface Water Response - AHF

Subsequent monitoring of AHF ponds after dosing indicated that tracer concentrations deep in the ponds remained significantly higher than concentrations in the shallower depths. Apparently, tracer leaving the pond with recharge water had to migrate upward and then horizontally to the sediment/ water interface in the shallow parts of the pond. Although no thermocline was present to indicate stratification, the vertical distribution of tracer suggests vertical mixing in the water column was incomplete, and the flux of tracer out of the pond may have been controlled by vertical diffusion at appreciable depths within the pond. Because of the relatively high elevation of the diversion pipes, surface water from Alameda Creek is received in the ponds at shallow depths beneath the ponds' water surface. In-flowing water may have percolated out of the ponds short-circuit-like fashion instead of mixing first with deeper pond water. Such a flow pattern would have increased the tracer residence time in the deeper parts of the ponds.

Surface Water Response- BHF

Incomplete vertical mixing in the Pit S portion of Lago Los Osos, as predicted through observation of a thermocline, was confirmed by the ^{22}Ne-tracer concentrations in surface water following dosing. In addition, the contrast in tracer concentrations among separate monitoring stations demonstrated that the remnants of the levee separating former Pits R and S act to resist complete lateral mixing throughout Lago Los Osos. However, some transfer of tracer mass did occur not only to the Pit R portion of Lago Los Osos but also to other ponds within the Quarry Pits complex.

Groundwater Response - AHF

The response of the AHF Aquifer to dosing of Kaiser and Bunting Ponds in the AHF sub-basin provided the most interesting results for this project. Tracers were detected in AHF monitoring wells rapidly after dosing of these two ponds. Except for well 21L07, tracer curves for the monitored wells have a steep upward curve to the peak, followed by a gradual downward slope that forms a long tail. These long tails are attributed at least partially to the long tracer residence time in the deep parts of the ponds. Unlike the other wells, well 21L07, located close to the inflow pipe, had a steep drop after the peak. The quick fall after the detection is attributed to possible flushing of a small embayment in the pond. Well 21L07 is located adjacent to this embayment.

Preferential pathways appear to have been responsible for the early detection of Kaiser and Bunting tracers at the PT Wellfield. These pathways could have been narrow-width channels or high conductivity layers, limited in thickness, but of possible broad lateral extent. This study provided evidence of both types of pathways. The fastest travel time was observed to be 7,000 feet/year, based on the initial detection of ^{136}Xe tracer (added to Kaiser Pond) in the PT Wellfield. The bulk of the tracer-tagged water moved more slowly, on time frames that may prove to be consistent with the water ages measured by the tritium-helium method. Evidence for such heterogeneity also includes the amount of dilution of the tracer signal in the wells and the difference between observed tracer arrival times and estimate given by the equation of Darcy's Law. The hydraulic conductivity used in the equation was determined by dividing a pump-test-derived transmissivity value by the full, saturated thickness of the aquifer.

All wells indicated slow mixing of tracer-laden water with water from different sources, or with water recharged during a time when tracer was not present. Moreover, mass balance considerations show that only a small fraction of the water recharged from AHF ponds reaches production wells within one year, further supporting the notion of heterogeneity in the AHF study area. It should be noted, however, that even if the transmissivity in the aquifer were uniform, some streamlines should take a slow, long circuitous route to the wellfield, while other streamlines should lead directly to the wellfield.

The AHF Alameda Creek experiment, for which ^{129}Xe was added upstream of Rubber Dam No. 3, was unsuccessful because a surface water transfer from the creek to adjacent recharge ponds could not be prevented. Hence, it was not possible to distinguish creek percolation from pond seepage. Detection of ^{129}Xe in wells located adjacent to AHF ponds were probably the result of this unintended surface water transfer.

Groundwater Response - BHF

Production wells in the Mowry Wellfield and several BHF monitoring wells were monitored over a ten-month period following the dosing of ^{22}Ne tracer in Pit S and Stevenson Pond in August of 1998. Tracer detections were limited to just two monitoring wells, both of which are screened in shallow intervals of the Newark Aquifer and located directly to the south of Pit S. There were no detections of tracer in the Mowry Wellfield or in other monitoring wells between the dosed ponds and the wellfield. Therefore, it can be concluded that water movement from the ponds to the Mowry Wellfield does not take place on the same short time frame as in the AHF sub-basin. The source of groundwater feeding the two Mowry production wells screened in the Newark Aquifer may be seepage from Kaiser Pond-BHF, a small pond that is fed by Kaiser Pond-AHF. Xenon-136, added to Kaiser Pond-AHF, was observed in two Newark

Aquifer wells in the Mowry Wellfield. Some water from other BHF ponds, including the ones dosed with ^{22}Ne in this study, may reach the deeper-screened production wells in a circuitous route. The rest of the water in the BHF ponds flows to more distal portions of the groundwater basin outside the study area.

CHARACTERIZATION OF WATER QUALITY DURING TRACER EXPERIMENTS

Documentation of water chemistry changes during recharge and subsequent migration in the aquifer was attempted but not achieved with the desired level of precision. Water from Pit S did not reach most of the BHF wells in the time frame of the study; therefore, results of samples collected from these wells are not comparable for purposes of fulfilling the original objectives of this study. In the AHF sub-basin, tracers did reach the all the targeted wells in the study area, but at very dilute concentrations relative to those observed in the recharge ponds after dosing. Tracer-tagged water ultimately mixed with deeper, ambient groundwater.

Despite the shortcomings of the water quality investigation, some of the observed changes in water quality parameters are dramatic and consistent over the study area, suggesting that some generalizations can be made. Several improvements to water quality appear to take place during recharge, over short times and distances. These include an elimination of male-specific coliphage and reductions in levels of TOC, MTBE, and nitrate. The only significant degradation of water quality is an increase in some major elements that comprise total dissolved solids. Coupled processes of dissolution of carbonate minerals and oxidative decomposition of organic matter at the sediment/water interface are consistent with the observed changes in geochemical parameters.

RECOMMENDATIONS

Over the course of the study, several lessons were learned that could be used to improve experimental design and outcome for future, similar research projects. Surface water bodies tend to be poor candidates for delivering tracer that can be reliably quantified down-gradient because of complicated hydraulics, incomplete mixing and uneven rates of recharge. A better approach for examining flow through the aquifer would be to inject tracer at an injection well(s) near recharge areas. Monitoring points with the necessary characteristics to allow quantitative analysis of flow and changes in water quality were lacking in the present study. Specifically, nested monitoring wells with short, shallow screens, should be used to examine variations in flow velocity with depth, and to collect water samples that are representative of a single source and recharge time. Additionally, monitoring surface water for tracer over a longer period of time after dosing would have enabled a better accounting of the tracer mass over time. For example, a large percentage of the mass of ^{136}Xe tracer not recovered at the Peralta-Tyson Wellfield was probably distributed throughout the deep portions of the aquifer where groundwater moves slowly relative to the shallower intervals of the aquifer. However, some of the non-recovered tracer may have been retained in the deep portion of Kaiser Pond. Without long-term pond monitoring, it was not possible to discern the fraction of tracer introduced in the pond from the fraction that remained in the pond over time.

Improvements could also be made to the approach used in this study to characterize surface water and groundwater chemistry. Multiple samples over time should have been collected from the ponds and nearby wells to enable the generation of concentration-versus-time plots for comparisons with tracer breakthrough curves. If a short-term dynamic approach timed with tracer experiments is determined to be too costly, then a long-term, quasi-steady state analysis may be a good alternative. For the latter, samples would need to be collected for at least four times per year and for multiple years to ensure observation of seasonal, as well as annual, variations in the water quality parameters of interest.

For the Alameda County Water District, this study has shown that AHF surface water supplies in particular need to be protected from contamination to the greatest extent possible, given the relatively fast travel time for from the ponds to the production wells. BHF production wells are not influenced by recharge water from the Quarry Pits complex ponds on a short (year) time frame; however, the travel time from the smaller Kaiser Pond-BHF to the wells is less than one year.

Results presented here can also be used to make informed decisions regarding pond operations and effects of recharge on water quality. For example, scraping sediments from the recharge ponds will undoubtedly lead to faster recharge, but may compromise the observed improvements in water quality that take place as water moves through a chemically reducing environment. Dissolution of aquifer minerals should be considered given the possibility of structural changes to the sediments near the sediment/water interface.

The results presented here can be used to develop hydrogeologic models for the portions of the two sub-basins within this study area. For example, particle tracking of water that originates in AHF ponds and is extracted at PT production wells will indicate a travel time that can be compared to both tracer travel times and groundwater ages. The model would have to couple surface water and groundwater processes, and handle three-dimensional groundwater flow. The model input would need to be structured to read spatially variant (vertically as well as horizontally) hydraulic conductivity values for the pond sediments as well as the aquifer. Pond sediment conductivity may also need to be time-dependent to handle changes over wetting and drying cycles. In addition, the surface water routine would have to account for incomplete vertical mixing. Incorporation of routines to model geochemical processes is also recommended. Such an effort may help to better identify the relative importance of the various processes at work in this study area. Overall, the model could be applied to guide possible future research in this study area, as well as to improve the interpretations of the results from this study.

Future research in this study area could include continued monitoring of tracers that were introduced during this investigation, and collecting additional data to refine estimates of groundwater ages. In addition, water quality changes through pond sediment, which were inferred in this study by simple comparison of surface water and groundwater chemistry, could be confirmed by high-resolution profiling of concentrations of constituents in the sediment layer. Dissolved phase concentrations and adsorbed phase concentrations in the sediment should be discerned. The sediment layers should also be characterized for permeability, thickness, and organic content. The range of targeted pond water constituents could be expanded to include humic substances, chlorophyll and algae, which would help explain nitrogen cycling and the source of TOC. Regular monitoring of an upstream station in Alameda Creek would help quantify the amount of constituent loads received from the Alameda Creek watershed upstream of ACWD.

CHAPTER 6
RECOMMENDATIONS TO UTILITIES

Use of age-dating and tracer techniques in combination with water quality characterization, as demonstrated in this study, can offer a number of potential benefits to water utilities whose responsibilities include management of groundwater basins and/or operation of municipal wells. The benefits include improved characterization of groundwater flow patterns, improved understanding of recharge pond seepage and hydrodynamics, improved understanding of geochemical processes and sources of recharge, opportunity for more reliable application of groundwater flow and water quality models, and more reliable vulnerability assessments. Utilities can also benefit from lessons learned through successes and shortcomings showcased in this study.

IMPROVING CHARACTERIZATION OF FLOW PATTERNS/ MODELING

Model-simulations of groundwater solute transport are sensitive to the distribution of hydraulic conductivity, aquifer thickness, and porosity within the modeled flow field. A combination of drilling logs, pump tests, and calibration is the conventional methodology for developing the required model input of aquifer properties. However, the difficulty of calibration increases with complexity of the flow domain that may include near-well environments, faults, streams, ponds, or multiple pumping wells. A three dimensional model may be required, further increasing the difficulty of calibration, if the aquifer layer is deep, aquifer properties change with depth, or if the wells only partially penetrate the aquifer.

Tracer study results can greatly improve calibration, and hence reliability, of a groundwater flow and solute-transport model (Thompson et al. 1999). As a minimum, a tracer study should enable determination of travel times between a point of injection and the receptor well with a high level of confidence. With multiple monitoring points, the tracer experiments can reveal heterogeneities. Modeling tracer results can help in the estimation of flow and transport parameters in complex flow regimes for which simple, direct calculations would not be adequate. If the utility's future operations are expected to be similar to pumping and recharge conditions experienced during the tracer study, then modeling may not be needed to realize the benefits of tracer study.

Noble gas isotopes can function advantageously as artificial tracers because they are inert, non-toxic, and can be detected even when greatly diluted. Because the isotopes used as tracers are low in abundance in natural waters, only a relatively small amount needs to be injected into the aquifer or dosed into a recharge pond. The potential for volatilization, however, needs to be recognized as a potential limitation of gas tracers, especially when applied to surface waters exposed to the atmosphere.

Because groundwater movement is typically slow, an appreciable amount of time may be required for a tracer experiment to yield useful results. Consequently, tracers may not be practical for mapping out wellhead protection zones coinciding with times of travel in excess of one or two years. In such cases, determining the age of the water may be an attractive alternative for inferring average travel times between recharge sources and production wells. Age dating data can also facilitate the calibration of models. Groundwater age (since the time of recharge) can be determined by the tritium-helium method described in Chapter 3 of this report. It should be pointed out that the water column in a pumping production well is a conglomeration of water particles pulled in from different radial directions in the aquifer. These particles of water can therefore have unique ages, and the statistical variation among samples from the same well can be large. Hence, characterization of preferential pathways toward the well should involve sampling from different monitoring points located strategically within the aquifer. Although ages can theoretically be calculated with an error of 0.1 years, a much lesser degree of precision was achieved in this study. The results of this project support the use of tracers as the preferred methodology for determining travel times in cases where travel times between recharge sources and production wells are likely to be two years or less. Nevertheless, utilities contemplating a tracer study should strongly consider age dating as a preliminary step to fine-tune the limits of the study area and to plan appropriate locations of monitoring wells and monitoring frequency.

POND HYDRODYNAMICS AND SEEPAGE

An unexpected benefit of this study was the demonstrated potential usefulness of isotopic tracers in studying pond hydrodynamics. The tracer experiments indicated the ease of dosing and monitoring tracers at various stations and depths within the ponds. The ease at detecting isotopic tracers at very low levels could make isotopic tracers attractive for this purpose. The study also has demonstrated that tracers could be beneficial in efforts to identify relative rates of seepage among various locations along the bottom and sides of the pond. However, such an investigation would probably require a more elaborate monitoring well network than what was available for this study. Ideally, clusters of monitoring wells should surround the pond, and each cluster should consist of multiple wells, with each well monitoring a specific depth within the aquifer. Tracer data could provide the basis of model calibration that, in turn, could lead to estimates of (possibly time-variant) dispersion coefficients controlling mass transfer within the pond.

Characterization of pond dynamics and seepage could enable recharge pond managers to better target pond maintenance efforts (such as dredging to improve permeability) and would assist greatly in the development of a reliable flow and/or water quality model that includes surface water-groundwater interactions.

Volatilization losses in surface waters open to the atmosphere can be substantial for dissolved gas tracers, and therefore, must be considered when estimating the mass of tracer fluxed through seepage or distributed within various parts of the pond. This limitation could also be a benefit in certain applications due to possible opportunity to estimate or confirm volatilization parameter values.

UNDERSTANDING GEOCHEMICAL PROCESSES AND DISTINGUISHING SOURCE WATERS

As demonstrated in this study (Chapter 4), analyzing water samples for natural isotopes (e.g., Carbon-13 and Carbon-14), in combination with conventional geochemical parameters (general minerals, trace metals, total organic carbon, pH, dissolved oxygen, oxidation-reduction potential, etc.) can lead to a better understanding of natural chemical processes that may alter water quality during recharge or during groundwater transport. Geochemical interpretation and/or modeling can help shed additional light on processes controlling local groundwater chemistry, such as precipitation-dissolution, complexation, and oxidation-reduction. Improved understanding of chemical processes in the local aquifer may be useful to water managers concerned with high levels of hardness, calcium, iron, manganese, sulfate, and TDS in production wells.

If the supply of recharge water is derived from multiple sources (i.e., local sources versus sources imported from locations many miles away), utilities may be able to distinguish the fate of the different recharge water sources by including Oxygen-18 and deuterium (*D*) among natural isotopes targeted for analysis. For utilities with large allotments of imported water, ^{18}O can be an excellent groundwater tracer, since imported sources usually differ greatly from local groundwater with respect to ^{18}O content. Accordingly, the ^{18}O-*D* data point for a local source will tend to have a noticeably different location along the Meteoric Water Line (MWL) than an imported source. Moreover, the departure of ^{18}O-*D* data points from the MWL can be interpreted to infer evaporation and/or certain geochemical processes to which the target surface water or groundwater has been subjected. For this study, these stable isotopes did not prove to be useful delineators of the extent of subsurface migration of artificial recharge because of the high degree of mixing of local and imported waters in the recharge ponds, as indicated by the ^{18}O-*D* data.

ASSESSING VULNERABILITY AND RESILIENCY

Vulnerability assessment for municipal wells is expected to become more common as a result of regulatory concerns with MTBE and pathogens. Multiple factors can contribute to a well's vulnerability, including (but not limited to) its construction, pumping rates, proximity to known contamination sources such as underground storage tanks, and proximity to surface water.

Water utilities operating wells and/or artificial recharge ponds in urban environments face potential hazards from chemical releases such as spills and leaking underground storage tanks. It is not unusual for service stations, dry cleaners, or trucking companies to be located close to municipal groundwater facilities. In response, wellhead protection programs are being implemented to ensure protection of the public's drinking water. Wellhead protection zones are commonly delineated by fixed or modified-fixed radius methods, or with the aid of numerical modeling. However, tracer studies can provide a more direct and reliable method for determining times of travel to the well, overcoming uncertainties inherent in conventional time of travel estimates.

Once the travel time between the production well and the potential contaminant source is established, utilities should be able to take appropriate precautionary measures to protect groundwater supplies, including determining the frequency of groundwater monitoring in the most sensitive portions of the wellhead protection zone. For example, ACWD has already used results of this tracer study to successfully negotiate a preventative monitoring plan with an oil company that wholesales gasoline to a service station located near its Peralta-Tyson Wellfield. Even though groundwater under the service station has been clean (and was so as of the date this report was written), the oil-company voluntarily agreed to sample two monitoring wells on a monthly basis for gasoline constituents including MTBE. Such frequent monitoring should provide early detection in the event of a leak, maximizing the opportunity to contain a plume before it reaches the wellfield.

In addition to potential chemical sources of contamination, such as leaking underground fuel tanks, water age and proximity to surface waters that recharge the aquifer have also been argued to be important risk factors for municipal wells. In particular, wells with young water are thought to be at greater risk than wells with older water as shorter residence times in porous media could mean that pathogens could reach the well instead of being immobilized en route. This risk would assume even greater significance when the source of recharge water is treated wastewater. The age-dating work and tracer experiments performed in this study have provided ACWD with a better sense of vulnerability of its wells to potential contamination. The water age of wells closest to the recharge ponds in the Above Hayward Fault Aquifer was approximately 4 years. However, tracer experiments demonstrate that a conservative solute introduced into the recharge ponds could reach at least one of these wells in less than 70 days. This contrast clearly shows that estimating travel time along the fastest flow paths to the well cannot necessarily be accomplished by age dating alone, or by input of pump test-derived values of hydraulic conductivity into the equation of Darcy's Law.

In this study, tracer travel times underscored vulnerability based on the fastest pathways to the well. On the other hand, a high degree of dilution and a long *average* travel time to the well, as were indicated in this study through the combined effort of tracer experiments and age dating, can be considered as measures of resiliency. In addition, adsorption and/or transformation in pond sediment or aquifer media may offer resiliency against certain undesirable non-conservative constituents, as indicated in this study with reductions in nitrate, TOC, and MTBE. The absence of detectable coliphage in groundwater samples indicates a potential capacity of the clogging layers in ponds to immobilize viral particles. However, the limited number of samples collected in this study (one to two per well and per pond), while providing very insightful information for ACWD, did not provide a basis for strong conclusions or firm estimates of transformation and retardation parameters. A proper assessment for such purposes would have required more than the reconnaissance level of effort performed in this study.

Timing a water chemistry monitoring program with a tracer study offers a rare opportunity to discern, through a short-term dynamic approach, transformation and adsorption processes from advection and dispersion, and thus provides a basis to estimate transformation and retardation parameters that can be used in a numeric model. As discussed in Chapter 4, concentration-versus-time plots would be needed for each target constituent in each recharge (surface water) source and in each of the wells in the monitoring network. Monitoring for the various constituents would need to span an appropriate length of time to assess background conditions prior to dosing as well as the time period after dosing for the tracers to break through. But such a level of effort may not be worthwhile unless there is an adequate stress in the system that occurs just as the tracer is introduced, for example, a sudden change in source water quality with a large increase in runoff during a storm event.

The observed differences between pond water and groundwater quality with respect to cations, alkalinity, pH, nitrate, and MTBE did offer some useful contrast for comparison in this study. And although there was some contrast in coliphage concentrations, the difference was less than 10 pfu per 100 ml, hardly a basis for a firm conclusion. Researchers that have investigated the mechanics of viral and bacterial transport in porous media under laboratory or field conditions have typically ensured an adequate stress by artificially injecting phage or bacteria (e.g., Bales et al. 1993, Bales et al. 1995, Jin et al. 1997, Scholl and Harvey 1992, Woessner et al. 2001). In many such experiments, the progress of phage migration has been compared with that of a tracer, and factors affecting phage adsorption or de-sorption have been investigated by adding solutions to alter pH or ionic strength.

A similarly intensive, highly focused experiment was beyond the scope of this study and probably would be too complex for most utilities managing groundwater basins. But other actions can be taken by utilities concerned about pathogens in groundwater, including sampling wells and recharge ponds, assessing potential sources of fecal contamination, and determining groundwater ages and travel times to wells. In addition, site-specific conditions can be compared with certain factors (soil properties, pH, ionic strength, etc.) that affect pathogen mobility in porous media, as previously revealed by researchers (such as those identified above or in Chapter 1) who have conducted highly focused and elaborate experiments or studies.

If investigating non-conservative behavior of various water quality parameters through a dynamic approach involving tracers is not practical, then comparing long-term trends in water quality could serve as a more attractive alternative for a utility attempting to assess transformation processes in aquifers and recharge ponds. Ultimately, efforts to protect groundwater supplies from gradual degradation in water quality will need to include protecting watersheds from which the source water is derived.

APPENDIX A
TABLES OF WATER LEVELS

Table A-1

Surface water elevations of selected recharge facilities (feet MSL)

DATE	Recharge Pond Elevations					Alameda Creek Elevations					
	Bunting	Kaiser (AHF)	Kaiser (BHF)	Pit S	Pit T1 & T2	Rubber Dam 1 Level	Rubber Dam 1 Up/Dn*	Rubber Dam 2 Level	Rubber Dam 2 Up/Dn*	Rubber Dam 3 Level	Rubber Dam 3 Up/Dn*
12/2/1997	59.3	45.5	45.5	-	15.2	42.8	2	29.8	2	52.7	2
12/16/1997	62.8	50.1	50.1	-	19.5	50.2	1	33.3	1	65.8	1
12/23/1997	62.9	51.1	51.1	-	22.5	51.1	1	33.3	1	65.0	1
12/30/1997	62.8	50.8	50.8	-	23.3	50.7	1	33.0	1	59.2	1
1/6/1998	61.5	49.7	49.7	14.6	20.8	50.4	1	33.3	1	52.7	2
1/13/1998	60.7	49.6	49.6	17.8	22.3	42.8	2	29.8	2	52.7	2
1/20/1998	60.0	48.2	48.2	17.2	19.7	42.8	2	29.8	2	52.7	2
1/27/1998	59.4	47.8	47.8	18.6	20.3	47.3	1	33.0	1	52.7	2
2/4/1998	61.5	49.0	49.0	23.4	21.1	45.7	2	32.4	2	54.6	2
2/17/1998	60.1	47.4	47.4	21.9	20.1	45.6	2	32.7	2	55.0	2
2/18/1998	60.0	47.3	47.3	21.8	20.0	45.4	2	32.5	2	54.8	2
2/24/1998	59.5	46.8	46.6	21.5	19.9	45.7	2	33.0	2	55.3	2
2/25/1998	59.4	46.7	46.5	21.4	19.8	45.5	2	32.6	2	55.0	2
3/3/1998	58.8	48.8	48.8	24.4	24.4	49.3	1	33.7	1	52.7	2
3/10/1998	58.2	50.8	50.8	29.9	23.9	51.0	1	33.0	1	52.7	2
3/17/1998	-	50.7	50.7	32.0	24.4	50.5	1	34.0	1	52.7	2
3/24/1998	B/S†	50.8	50.8	32.8	25.1	42.8	2	29.8	2	54.0	2
3/31/1998	B/S†	49.0	49.0	31.7	25.2	42.8	2	29.8	2	52.7	2
4/7/1998	B/S†	48.1	48.1	29.9	25.1	44.0	2	29.9	2	52.7	2
4/8/1998	B/S†	48.1	48.1	29.7	25.1	43.9	2	29.8	2	52.7	2
4/21/1998	-	51.2	51.2	33.0	25.0	51.3	1	33.5	1	52.7	2
4/28/1998	-	50.6	50.6	33.3	25.6	50.6	1	33.5	1	52.7	2
5/5/1998	59.1	50.7	50.7	33.9	26.0	51.4	1	33.7	1	65.3	2
5/12/1998	61.4	50.7	50.7	33.3	26.3	42.8	2	29.8	2	52.7	2
5/19/1998	58.4	50.8	50.8	33.0	26.3	51.4	1	33.0	1	57.3	1
5/26/1998	63.3	50.0	50.0	34.0	26.5	50.0	1	32.9	1	65.3	1
6/2/1998	60.6	51.5	51.5	-	26.7	52.0	1	34.2	1	52.7	2
6/9/1998	58.9	50.0	50.0	34.1	26.9	-	-	-	-	52.7	2
6/16/1998	-	50.0	50.0	33.9	26.8	50.1	1	33.9	1	52.7	2
6/23/1998	-	51.0	51.0	33.1	26.7	51.3	1	33.7	1	52.7	2
6/30/1998	61.3	51.0	51.0	31.8	26.5	51.0	1	33.5	1	65.5	1
7/7/1998	63.2	47.7	47.7	32.4	26.3	42.8	2	33.4	1	65.4	1
7/14/1998	63.2	49.0	49.0	31.2	26.1	49.6	1	33.3	1	65.6	1
7/21/1998	62.9	50.2	50.2	29.7	25.6	50.4	1	33.2	1	65.5	1
7/28/1998	63.1	50.5	50.5	29.0	25.7	50.5	1	33.0	1	65.4	1
8/13/1998	62.1	46.2	46.2	29.2	26.1	48.2	1	32.9	1	62.5	1
8/18/1998	61.9	48.4	47.2	28.5	26.9	48.8	1	33.0	1	62.6	1
8/24/1998	60.8	48.8	48.8	27.7	26.0	49.0	1	32.4	1	62.9	1
8/25/1998	60.5	49.0	49.0	27.5	25.8	49.1	1	32.4	1	62.4	1
9/2/1998	61.3	48.2	48.2	26.7	25.6	48.2	1	32.4	1	64.0	1
9/8/1998	62.9	48.2	48.2	25.9	24.4	49.1	1	32.2	1	65.7	1
9/10/1998	63.0	49.0	49.0	25.7	24.1	49.4	1	32.3	1	65.1	1
9/15/1998	63.2	50.0	50.0	25.0	24.5	50.1	1	33.0	1	65.7	1
9/22/1998	63.2	50.7	50.7	24.6	25.7	50.8	2	32.9	1	65.8	1
9/29/1998	63.3	51.2	51.2	24.2	27.0	51.2	1	33.4	1	65.6	1
10/2/1998	63.3	51.2	51.2	24.0	25.8	51.5	1	32.0	1	65.6	1
10/6/1998	63.2	51.3	51.3	-	25.0	51.3	1	32.3	1	65.7	1
10/13/1998	63.2	50.6	50.6	23.5	25.2	50.6	1	32.9	1	65.8	1

(continued)

Table A-1 (continued)

DATE	Recharge Pond Elevations Bunting	Kaiser (AHF)	Kaiser (BHF)	Pit S	Pit T1 & T2	Alameda Creek Elevations Rubber Dam 1 Level	Rubber Dam 1 Up/Dn*	Rubber Dam 2 Level	Rubber Dam 2 Up/Dn*	Rubber Dam 3 Level	Rubber Dam 3 Up/Dn*
10/20/1998	63.1	50.2	50.2	23.0	24.4	50.3	1	32.3	1	65.7	1
10/28/1998	62.9	51.0	51.0	22.6	23.0	51.8	1	34.6	1	65.7	1
11/3/1998	63.3	51.0	51.0	22.3	23.4	51.2	1	33.4	1	65.7	1
11/9/1998	62.8	50.1	50.1	22.0	23.7	49.5	1	33.6	1	52.7	2
11/17/1998	63.0	49.3	49.3	22.2	23.7	49.3	1	32.7	1	65.7	1
11/23/1998	62.1	48.3	48.3	24.0	23.6	45.8	1	31.9	1	52.7	2
12/1/1998	60.8	47.2	47.2	25.4	23.2	42.8	2	29.8	2	52.7	2
12/8/1998	59.9	49.9	49.9	24.3	23.5	50.5	1	33.5	1	52.7	2
12/16/1998	62.5	51.0	51.0	26.2	24.9	51.0	1	32.7	1	65.5	1
12/29/1998	62.6	51.3	51.3	29.5	25.1	51.3	1	32.8	1	65.7	1
1/5/1999	62.9	51.7	51.7	29.4	24.6	51.8	1	34.4	1	65.6	1
1/12/1999	62.8	50.9	50.9	29.1	24.2	50.9	1	34.0	1	65.6	1
1/20/1999	62.0	49.4	49.4	30.0	24.4	44.5	2	29.8	2	52.7	2
1/27/1999	61.1	49.0	49.0	31.3	24.4	49.5	1	33.2	1	52.7	2
2/2/1999	60.5	51.0	51.0	32.6	24.8	51.0	1	32.5	1	52.7	2
2/16/1999	63.0	50.5	50.5	34.5	25.6	42.8	2	29.8	2	65.4	1
2/23/1999	61.9	48.9	48.9	33.2	26.0	42.8	2	32.8	1	52.7	2
3/11/1999	62.3	50.7	50.7	33.7	26.3	47.0	1	34.4	1	52.7	2
3/31/1999	60.7	51.4	51.4	34.0	26.7	51.5	1	34.7	1	52.7	2
4/7/1999	62.7	50.6	50.6	33.9	26.6	42.8	2	29.8	2	52.7	2
4/14/1999	61.5	49.0	49.0	32.6	26.4	51.0	1	34.7	1	52.7	2
4/20/1999	63.2	51.4	51.4	34.8	26.6	51.4	1	34.9	1	65.7	1
4/29/1999	63.2	51.5	51.5	34.4	26.9	51.5	1	34.9	1	65.5	1
5/4/1999	63.3	51.5	51.5	33.5	27.0	51.5	1	35.1	1	65.4	1
5/11/1999	62.9	51.5	51.5	32.5	26.9	51.5	1	35.0	1	65.6	1
5/18/1999	63.0	51.4	51.4	31.4	26.6	51.4	1	35.0	1	65.5	1
5/26/1999	62.9	51.3	51.3	30.4	26.2	51.3	1	35.0	1	65.6	1
6/1/1999	62.9	51.6	51.6	29.7	25.9	51.7	1	34.9	1	65.5	1
6/15/1999	63.0	51.1	51.1	28.4	25.0	51.1	1	33.7	1	65.6	1
6/22/1999	61.4	50.6	50.6	28.5	24.7	50.6	1	34.4	1	57.5	1

* Up/Dn signifies whether the rubber dam was up (1) or down (2) at the time of the recording.

Table A-2

Groundwater piezometric head (feet MSL)

DATE	Above Hayward Fault																					
	PT 8		21L03		21L05		21J27		21J03		21K05		21L08		21L06		21L07		27D08		21H11	
	MPE* = 69.025		MPE= 66.565		MPE= 67.460		MPE= 74.480		MPE= 77.300		MPE= 69.830		MPE= 66.940		MPE= 67.810		MPE= 70.615		MPE= 66.475		MPE= 75.680	
	DTW	Elev.	DTW	Elev.	DTW	Elev.	DTW	Elev.	DTW	Elev.	DTW	Elev.	DTW	Elev.	DTW	Elev.	DTW	Elev.	DTW	Elev.	DTW	Elev.
12/2/1997	-	-	38.10	28.47	38.70	28.76	-	-	-	-	-	-	-	-	39.00	28.81	47.00	23.62	42.20	24.28	-	-
12/16/1997	-	-	-	-	-	-	-	-	44.50	32.80	-	-	29.70	37.24	-	-	-	-	39.80	26.68	-	-
12/23/1997	-	-	44.50	22.07	-	-	-	-	-	-	-	-	-	-	-	-	-	-	38.20	28.28	-	-
12/30/1997	-	-	33.40	33.17	-	-	-	-	43.20	34.10	-	-	29.80	37.14	34.10	33.71	36.20	34.42	37.10	29.38	-	-
1/6/1998	-	-	34.90	31.67	-	-	-	-	43.30	34.00	-	-	33.80	33.14	34.40	33.41	-	-	36.10	30.38	-	-
1/13/1998	-	-	-	-	34.30	33.16	-	-	43.60	33.70	-	-	33.90	33.04	34.40	33.41	36.90	33.72	35.80	30.68	-	-
1/20/1998	-	-	-	-	35.30	32.16	-	-	44.40	32.90	37.70	32.13	34.90	32.04	35.30	32.51	37.60	33.02	35.60	30.88	-	-
1/27/1998	-	-	36.30	30.27	36.50	30.96	-	-	-	-	38.70	31.13	36.00	30.94	36.60	31.21	38.80	31.82	35.70	30.78	-	-
2/4/1998	-	-	34.30	32.27	34.50	32.96	-	-	-	-	37.00	32.83	-	-	34.50	33.31	36.70	33.92	-	-	-	-
2/10/1998	-	-	34.50	32.07	34.90	32.56	-	-	-	-	37.30	32.53	34.00	32.94	34.90	32.91	36.90	33.72	34.30	32.18	-	-
2/17/1998	-	-	34.20	32.37	34.50	32.96	-	-	43.00	34.30	-	-	34.20	32.74	34.70	33.11	36.60	34.02	33.60	32.88	-	-
2/18/1998	-	-	-	-	-	-	-	-	-	-	36.60	33.23	-	-	-	-	-	-	-	-	-	-
2/24/1998	-	-	34.00	32.57	-	-	41.20	33.28	42.60	34.70	36.50	33.33	-	-	-	-	-	-	33.00	33.48	-	-
2/25/1998	-	-	-	-	34.60	32.86	-	-	-	-	-	-	34.30	32.64	34.60	33.21	36.70	33.92	-	-	-	-
3/3/1998	-	-	34.10	32.47	-	-	41.50	32.98	43.40	33.90	-	-	-	-	-	-	-	-	32.80	33.68	-	-
3/10/1998	-	-	31.50	35.07	32.60	34.86	-	-	-	-	34.80	35.03	32.10	34.84	32.50	35.31	35.00	35.62	32.40	34.08	-	-
3/17/1998	-	-	30.10	36.47	-	-	-	-	28.90	48.40	-	-	29.70	37.24	-	-	31.80	38.82	30.80	35.68	-	-
3/24/1998	-	-	29.50	37.07	30.20	37.26	42.00	32.48	-	-	32.60	37.23	29.90	37.04	30.30	37.51	32.60	38.02	30.60	35.88	-	-
3/31/1998	-	-	29.60	36.97	-	-	37.30	37.18	39.50	37.80	-	-	-	-	-	-	-	-	29.90	36.58	-	-
4/7/1998	-	-	-	-	-	-	37.90	36.58	39.80	37.50	-	-	-	-	-	-	-	-	29.90	36.58	-	-
4/8/1998	-	-	-	-	31.50	35.96	-	-	-	-	33.50	36.33	31.30	35.64	31.60	36.21	33.90	36.72	-	-	-	-
4/21/1998	-	-	-	-	30.90	36.56	37.80	36.68	40.20	37.10	33.20	36.63	30.80	36.14	31.80	36.01	-	-	30.00	36.48	-	-
4/28/1998	-	-	-	-	30.90	36.56	38.10	36.38	-	-	33.30	36.53	-	-	-	-	-	-	30.00	36.48	-	-
5/5/1998	-	-	-	-	-	-	36.70	37.78	-	-	31.40	38.43	-	-	-	-	-	-	29.90	36.58	-	-
5/12/1998	-	-	34.90	31.67	26.90	40.56	-	-	-	-	-	-	26.90	40.04	26.80	41.01	-	-	28.90	37.58	-	-
5/19/1998	-	-	-	-	-	-	34.70	39.78	-	-	-	-	-	-	-	-	-	-	27.90	38.58	-	-
5/26/1998	-	-	-	-	26.70	40.76	-	-	34.70	42.60	28.40	41.43	-	-	26.60	41.21	-	-	28.00	38.48	-	-
6/2/1998	-	-	-	-	-	-	34.40	40.08	-	-		-	-	-	-	-	-	-	27.70	38.78	-	-
6/9/1998	33.14	35.89	27.95	38.62	28.94	38.52	35.58	38.90	37.58	39.72	30.83	39.00	28.84	38.10	28.90	38.91	31.70	38.92	27.85	38.63	-	-
6/16/1998	36.94	32.09	-	-	-	-	36.45	38.03	-	-	-	-	-	-	-	-	-	-	28.30	38.18	-	-
6/23/1998	37.61	31.42	-	-	-	-	-	-	-	-	-	-	-	-	-	-	-	-	29.60	36.88	-	-

(continued)

Table A-2 (continued)

DATE	Above Hayward Fault																					
	PT 8		21L03		21L05		21J27		21J03		21K05		21L08		21L06		21L07		27D08		21H11	
	MPE* = 69.025		MPE= 66.565		MPE= 67.460		MPE= 74.480		MPE= 77.300		MPE= 69.830		MPE= 66.940		MPE= 67.810		MPE= 70.615		MPE= 66.475		MPE= 75.680	
	DTW	Elev.	DTW	Elev.	DTW	Elev.	DTW	Elev.	DTW	Elev.	DTW	Elev.	DTW	Elev.	DTW	Elev.	DTW	Elev.	DTW	Elev.	DTW	Elev.
6/30/1998	38.87	30.16	32.82	33.75	33.16	34.30	39.55	34.93	40.46	36.84	34.87	34.96	32.80	34.14	33.00	34.81	-	-	30.69	35.79	-	-
7/7/1998	35.00	34.03	29.25	37.32	29.81	37.65	35.98	38.50	36.88	40.42	31.15	38.68	29.62	37.32	-	-	-	-	29.80	36.68	-	-
7/14/1998	37.00	32.03	30.83	35.74	31.28	36.18	37.70	36.78	32.88	44.42	38.74	31.09	31.11	35.83	31.12	36.69	32.94	37.68	30.30	36.18	-	-
7/21/1998	37.57	31.46	31.58	34.99	32.24	35.22	39.04	35.44	34.05	43.25	34.11	35.72	31.95	34.99	32.16	35.65	34.01	36.61	31.00	35.48	-	-
7/28/1998	40.79	28.24	33.95	32.62	33.31	34.15	40.05	34.43	40.95	36.35	35.25	34.58	32.92	34.02	33.22	34.59	34.97	35.65	31.60	34.88	-	-
8/13/1998	43.75	25.28	36.61	29.96	35.67	31.79	41.68	32.80	43.11	34.19	37.40	32.43	35.26	31.68	35.67	32.14	37.62	33.00	-	-	-	-
8/18/1998	48.63	20.40	42.83	23.74	42.13	25.33	48.27	26.21	49.68	27.62	43.96	25.87	40.30	26.64	42.51	25.30	44.03	26.59	36.40	30.08	-	-
8/24/1998	-	-	-	-	-	-	-	-	-	-	-	-	-	-	-	-	-	-	-	-	-	-
8/25/1998	50.10	18.93	43.66	22.91	42.76	24.70	49.45	25.03	50.82	26.48	44.92	24.91	41.95	24.99	42.81	25.00	44.63	25.99	37.70	28.78	40.92	34.76
9/1/1998	38.87	30.16	-	-	-	-	-	-	50.14	27.16	-	-	-	-	-	-	-	-	38.30	28.18	41.08	34.60
9/2/1998	36.88	32.15	41.00	25.57	41.17	26.29	48.20	26.28	-	-	43.40	26.43	40.61	26.33	41.27	26.54	43.29	27.33	-	-	-	-
9/8/1998	-	-	-	-	-	-	-	-	-	-	-	-	-	-	-	-	-	-	38.40	28.08	-	-
9/10/1998	-	-	-	-	-	-	-	-	-	-	-	-	-	-	-	-	-	-	-	-	-	-
9/15/1998	43.38	25.65	36.60	29.97	36.58	30.88	44.32	30.16	44.28	33.02	38.75	31.08	31.51	35.43	36.56	31.25	38.25	32.37	37.70	28.78	33.30	42.38
9/22/1998	40.91	28.12	35.25	31.32	35.76	31.70	44.11	30.37	44.46	32.84	38.62	31.21	30.15	36.79	35.77	32.04	37.50	33.12	37.60	28.88	33.06	42.62
9/29/1998	39.82	29.21	33.85	32.72	-	-	-	-	42.80	34.50	36.65	33.18	-	-	-	-	-	-	36.80	29.68	32.32	43.36
10/2/1998	-	-	-	-	-	-	-	-	-	-	-	-	-	-	-	-	-	-	-	-	-	-
10/6/1998	41.48	27.55	34.68	31.89	34.72	32.74	42.15	32.33	-	-	-	-	30.04	36.90	34.71	33.10	36.47	34.15	-	-	32.47	43.21
10/13/1998	43.73	25.30	36.48	30.09	36.40	31.06	43.72	30.76	44.08	33.22	38.42	31.41	36.02	30.92	36.32	31.49	38.24	32.38	36.50	29.98	-	-
10/20/1998	45.97	23.06	38.31	28.26	37.92	29.54	45.31	29.17	45.55	31.75	40.32	29.51	37.70	29.24	37.86	29.95	39.83	30.79	37.00	29.48	35.00	40.68
10/28/1998	45.60	23.43	38.37	28.20	37.40	30.06	45.67	28.81	45.91	31.39	39.78	30.05	36.65	30.29	37.35	30.46	39.25	31.37	37.60	28.88	36.02	39.66
11/3/1998	44.28	24.75	38.40	28.17	38.15	29.31	45.40	29.08	46.00	31.30	40.27	29.56	37.26	29.68	38.13	29.68	39.85	30.77	37.70	28.78	35.99	39.69
11/9/1998	42.06	26.97	37.37	29.20	38.06	29.40	45.72	28.76	-	-	-	-	-	-	37.99	29.82	40.25	30.37	37.90	28.58	38.49	37.19
11/17/1998	42.17	26.86	37.14	29.43	37.14	30.32	44.90	29.58	46.46	30.84	40.10	29.73	36.71	30.23	37.62	30.19	39.45	31.17	37.70	28.78	-	-
11/23/1998	39.21	29.82	35.43	31.14	36.00	31.46	43.61	30.87	-	-	-	-	35.64	31.30	36.32	31.49	38.47	32.15	37.80	28.68	-	-
12/1/1998	38.77	30.26	34.53	32.04	34.95	32.51	42.00	32.48	44.43	32.87	37.50	32.33	-	-	35.09	32.72	-	-	34.80	31.68	-	-
12/8/1998	39.46	29.57	35.25	31.32	35.82	31.64	43.17	31.31	45.00	32.30	38.37	31.46	35.02	31.92	35.92	31.89	38.31	32.31	35.80	30.68	40.02	35.66
12/16/1998	36.63	32.40	32.31	34.26	33.00	34.46	40.85	33.63	-	-	-	-	30.00	36.94	33.22	34.59	35.49	35.13	34.90	31.58	33.48	42.20
12/29/1998	33.23	35.80	28.70	37.87	29.51	37.95	37.38	37.10	38.90	38.40	32.08	37.75	29.39	37.55	29.56	38.25	31.76	38.86	32.20	34.28	29.92	45.76
1/5/1999	-	-	-	-	-	-	-	-	-	-	-	-	-	-	-	-	-	-	31.50	34.98	-	-
1/12/1999	32.95	36.08	-	-	29.50	37.96	36.83	37.65	-	-	-	-	29.35	37.59	29.53	38.28	31.73	38.89	30.70	35.78	-	-
1/20/1999	33.00	36.03	29.41	37.16	30.65	36.81	37.21	37.27	39.27	38.03	32.21	37.62	30.50	36.44	30.69	37.12	33.00	37.62	-	-	-	-
1/27/1999	31.28	37.75	-	-	-	-	36.55	37.93	38.88	38.42	31.62	38.21	-	-	-	-	-	-	29.90	36.58	34.52	41.16

(continued)

Table A-2 (continued)

DATE	Above Hayward Fault																					
	PT 8		21L03		21L05		21J27		21J03		21K05		21L08		21L06		21L07		27D08		21H11	
	MPE* = 69.025		MPE= 66.565		MPE= 67.460		MPE= 74.480		MPE= 77.300		MPE= 69.830		MPE= 66.940		MPE= 67.810		MPE= 70.615		MPE= 66.475		MPE= 75.680	
	DTW	Elev.	DTW	Elev.	DTW	Elev.	DTW	Elev.	DTW	Elev.	DTW	Elev.	DTW	Elev.	DTW	Elev.	DTW	Elev.	DTW	Elev.	DTW	Elev.
2/2/1999	31.35	37.68	28.48	38.09	29.72	37.74	29.45	45.03	38.60	38.70	31.62	38.21	-	-	29.72	38.09	32.10	38.52	29.10	37.38	33.96	41.72
2/16/1999	32.64	36.39	-	-	29.51	37.95	36.42	38.06	38.14	39.16	-	-	29.35	37.59	29.50	38.31	31.75	38.87	28.80	37.68	30.33	45.35
2/23/1999	31.10	37.93	27.86	38.71	28.95	38.51	36.00	38.48	38.15	39.15	31.60	38.23	28.91	38.03	29.10	38.71	31.46	39.16	28.30	38.18	33.11	42.57
3/11/1999	-	-	27.57	39.00	-	-	34.42	40.06	36.56	40.74	29.68	40.15	27.53	39.41	-	-	29.86	40.76	27.40	39.08	30.18	45.50
3/31/1999	29.30	39.73	26.12	40.45	27.51	39.95	34.50	39.98	36.88	40.42	-	-	27.57	39.37	27.61	40.20	30.02	40.60	26.70	39.78	32.45	43.23
4/7/1999	29.18	39.85	-	-	-	-	-	-	-	-	-	-	-	-	-	-	-	-	-	-	-	-
4/14/1999	30.74	38.29	26.78	39.79	27.76	39.70	29.48	45.00	36.48	40.82	-	-	27.82	39.12	27.81	40.00	30.16	40.46	25.50	40.98	-	-
4/20/1999	-	-	28.10	38.47	28.40	39.06	34.82	39.66	-	-	-	-	28.20	38.74	28.50	39.31	30.40	40.22	25.40	41.08	27.40	48.28
4/29/1999	-	-	30.80	35.77	30.18	37.28	36.75	37.73	-	-	-	-	29.94	37.00	30.14	37.67	32.08	38.54	28.20	38.28	28.58	47.10
5/4/1999	-	-	35.00	31.57	33.89	33.57	39.59	34.89	-	-	-	-	-	-	33.81	34.00	35.54	35.08	-	-	29.97	45.71
5/11/1999	-	-	36.01	30.56	34.78	32.68	41.56	32.92	-	-	-	-	34.30	32.64	34.78	33.03	36.55	34.07	30.40	36.08	-	-
5/18/1999	-	-	38.91	27.66	37.80	29.66	44.25	30.23	45.14	32.16	39.74	30.09	37.19	29.75	37.70	30.11	39.33	31.29	33.30	33.18	34.45	41.23
5/26/1999	-	-	37.81	28.76	36.65	30.81	44.27	30.21	-	-	-	-	36.14	30.80	36.71	31.10	38.60	32.02	33.80	32.68	35.22	40.46
6/1/1999	-	-	37.94	28.63	37.55	29.91	44.28	30.20	45.28	32.02	-	-	36.95	29.99	37.52	30.29	39.23	31.39	35.40	31.08	35.38	40.30
6/15/1999	47.53	21.50	-	-	39.66	27.80	46.87	27.61	47.61	29.69	-	-	-	-	39.60	28.21	41.27	29.35	37.70	28.78	-	-
6/22/1999	-	-	42.11	24.46	40.83	26.63	48.17	26.31	48.92	28.38	43.21	26.62	40.10	26.84	40.78	27.03	42.43	28.19	38.50	27.98	38.69	36.99

DATE	Below Hayward Fault																					
	20J04		20J06		20J05		20R03		20R04		20R05		20R07		20R02		Mowry 8		28D12		28D11	
	MPE† = 58.18		MPE† = 58.41		MPE† = 58.47		MPE† = 57.88		MPE† = 57.84		MPE† = 57.97		MPE= 62.810		MPE= 65.364		MPE ‡ = 64.242		MPE= 62.860		MPE= 62.900	
	DTW	Elev.	DTW	Elev.	DTW	Elev.	DTW	Elev.	DTW	Elev.	DTW	Elev.	DTW	Elev.	DTW	Elev.	DTW	Elev.	DTW	Elev.	DTW	Elev.
8/24/1998	33.06	25.12	34.01	24.40	38.10	20.37	34.33	23.55	36.11	21.73	37.48	20.49	-	-	43.56	21.80	50.04	14.20	45.37	17.49	-	-
8/25/1998	33.06	25.12	34.01	24.40	-	-	34.33	23.55	-	-	-	-	-	-	-	-	-	-	-	-	-	-
9/1/1998	33.69	24.49	34.63	23.78	38.70	19.77	34.87	23.01	36.65	21.19	38.02	19.95	-	-	45.97	19.39	52.80	11.44	46.03	16.83	-	-
9/2/1998	-	-	-	-	-	-	-	-	-	-	-	-	-	-	-	-	-	-	-	-	-	-
9/8/1998	-	-	-	-	-	-	-	-	-	-	-	-	-	-	-	-	-	-	-	-	-	-
9/10/1998	-	-	-	-	-	-	-	-	-	-	-	-	40.11	22.70	-	-	-	-	-	-	-	-
9/15/1998	34.28	23.90	35.30	23.11	39.84	18.63	36.17	21.71	37.90	19.94	39.24	18.73	40.12	22.69	45.13	20.23	51.51	12.73	47.14	15.72	-	-

(continued)

Table A-2 (continued)

DATE	Below Hayward Fault																					
	20J04		20J06		20J05		20R03		20R04		20R05		20R07		20R02		Mowry 8		28D12		28D11	
	MPE† = 58.18		MPE† = 58.41		MPE† = 58.47		MPE† = 57.88		MPE† = 57.84		MPE† = 57.97		MPE= 62.810		MPE= 65.364		MPE ‡ = 64.242		MPE= 62.860		MPE= 62.900	
	DTW	Elev.	DTW	Elev.	DTW	Elev.	DTW	Elev.	DTW	Elev.	DTW	Elev.	DTW	Elev.	DTW	Elev.	DTW	Elev.	DTW	Elev.	DTW	Elev.
9/22/1998	33.97	24.21	35.04	23.37	39.76	18.71	35.97	21.91	37.77	20.07	39.15	18.82	40.00	22.81	45.19	20.17	52.15	12.09	47.13	15.73	-	-
9/29/1998	33.57	24.61	34.67	23.74	39.61	18.86	38.50	19.38	37.52	20.32	39.00	18.97	39.77	23.04	44.90	20.46	52.02	12.22	47.20	15.66	-	-
10/2/1998	-	-	-	-	-	-	38.50	19.38	-	-	-	-	-	-	-	-	-	-	-	-	-	-
10/6/1998	34.61	23.57	35.63	22.78	40.13	18.34	36.69	21.19	38.31	19.53	39.54	18.43	40.44	22.37	45.50	19.86	52.11	12.13	47.38	15.48	-	-
10/13/1998	35.34	22.84	36.31	22.10	40.55	17.92	37.04	20.84	38.63	19.21	39.90	18.07	40.74	22.07	45.88	19.48	52.00	12.24	47.41	15.45	-	-
10/20/1998	36.13	22.05	37.03	21.38	40.97	17.50	37.55	20.33	39.00	18.84	40.14	17.83	41.22	21.59	46.22	19.14	52.33	11.91	47.82	15.04	-	-
10/28/1998	35.20	22.98	36.21	22.20	41.12	17.35	37.28	20.60	39.14	18.70	40.53	17.44	41.29	21.52	46.45	18.91	52.32	11.92	47.67	15.19	-	-
11/3/1998	36.25	21.93	37.23	21.18	41.50	16.97	38.16	19.72	39.68	18.16	40.89	17.08	41.70	21.11	46.82	18.54	52.97	11.27	48.25	14.61	-	-
11/9/1998	37.35	20.83	38.18	20.23	41.63	16.84	38.63	19.25	39.80	18.04	41.06	16.91	42.22	20.59	46.80	18.56	53.10	11.14	48.41	14.45	-	-
11/17/1998	37.24	20.94	38.56	19.85	41.12	17.35	38.75	19.13	40.22	17.62	41.38	16.59	42.56	20.25	47.24	18.12	53.60	10.64	48.80	14.06	-	-
11/23/1998	37.35	20.83	38.47	19.94	41.64	16.83	38.28	19.60	39.28	18.56	41.00	16.97	42.58	20.23	46.94	18.42	48.70	15.54	48.04	14.82	58.70	4.20
12/1/1998	36.58	21.60	37.41	21.00	41.15	17.32	37.66	20.22	39.28	18.56	40.58	17.39	41.94	20.87	46.40	18.96	47.94	16.30	46.35	16.51	-	-
12/8/1998	36.73	21.45	37.50	20.91	41.13	17.34	37.67	20.21	39.23	18.61	40.49	17.48	41.87	20.94	46.22	19.14	52.15	12.09	46.00	16.86	55.68	7.22
12/16/1998	35.74	22.44	36.58	21.83	40.60	17.87	36.85	21.03	38.60	19.24	39.99	17.98	41.35	21.46	43.80	21.56	48.67	15.57	46.00	16.86	-	-
12/29/1998	33.14	25.04	34.08	24.33	39.00	19.47	34.82	23.06	37.00	20.84	38.70	19.27	39.24	23.57	44.50	20.86	47.15	17.09	45.55	17.31	-	-
1/5/1999	32.44	25.74	33.40	25.01	38.75	19.72	34.41	23.47	36.67	21.17	38.46	19.51	38.55	24.26	44.10	21.26	-	-	44.91	17.95	59.04	3.86
1/12/1999	-	-	-	-	-	-	34.78	23.10	36.95	20.89	38.63	19.34	38.56	24.25	44.50	20.86	51.64	12.60	46.47	16.39	-	-
1/20/1999	-	-	-	-	-	-	-	-	-	-	-	-	38.80	24.01	44.32	21.04	47.97	16.27	45.16	17.70	-	-
1/27/1999	-	-	-	-	-	-	-	-	-	-	-	-	38.51	24.30	43.70	21.66	47.35	16.89	44.49	18.37	-	-
2/2/1999	33.16	25.02	35.57	22.84	37.38	21.09	-	-	-	-	-	-	37.62	25.19	43.15	22.21	50.54	13.70	44.28	18.58	-	-
2/16/1999	-	-	-	-	-	-	-	-	-	-	-	-	36.75	26.06	42.19	23.17	47.75	16.49	44.52	18.34	-	-
2/23/1999	-	-	-	-	-	-	-	-	-	-	-	-	36.80	26.01	42.25	23.11	49.47	14.77	44.06	18.80	-	-
3/11/1999	-	-	-	-	-	-	-	-	-	-	-	-	35.95	26.86	41.50	23.86	44.45	19.79	42.06	20.80	49.47	13.43
3/31/1999	29.02	29.16	30.00	28.41	35.41	23.06	31.04	26.84	-	-	-	-	35.11	27.70	40.84	24.52	-	-	41.43	21.43	-	-
4/7/1999	-	-	-	-	-	-	-	-	-	-	-	-	35.18	27.63	41.00	24.36	47.23	17.01	42.69	20.17	-	-
4/14/1999	29.83	28.35	30.78	27.63	35.75	22.72	33.80	24.08	31.57	26.27	35.38	22.59	-	-	-	-	47.56	16.68	42.60	20.26	-	-
4/20/1999	28.50	29.68	29.50	28.91	35.10	23.37	30.50	27.38	33.00	24.84	34.80	23.17	35.10	27.71	40.90	24.46	47.29	16.95	42.39	20.47	49.10	13.80
4/29/1999	28.06	30.12	29.12	29.29	34.57	23.90	30.35	27.53	32.75	25.09	34.35	23.62	34.53	28.28	-	-	42.43	21.81	40.60	22.26	-	-
5/4/1999	28.30	29.88	29.30	29.11	34.61	23.86	-	-	-	-	-	-	34.50	28.31	40.10	25.26	42.58	21.66	40.58	22.28	-	-
5/11/1999	-	-	-	-	-	-	-	-	-	-	-	-	34.88	27.93	40.28	25.08	42.21	22.03	40.53	22.33	-	-

(continued)

Table A-2 (continued)

DATE	Below Hayward Fault																					
	20J04		20J06		20J05		20R03		20R04		20R05		20R07		20R02		Mowry 8		28D12		28D11	
	MPE†= 58.18		MPE†= 58.41		MPE†= 58.47		MPE†= 57.88		MPE†= 57.84		MPE†= 57.97		MPE= 62.810		MPE= 65.364		MPE ‡ = 64.242		MPE= 62.860		MPE= 62.900	
	DTW	Elev.	DTW	Elev.	DTW	Elev.	DTW	Elev.	DTW	Elev.	DTW	Elev.	DTW	Elev.	DTW	Elev.	DTW	Elev.	DTW	Elev.	DTW	Elev.
5/18/1999	29.45	28.73	30.50	27.91	35.39	23.08	31.47	26.41	33.57	24.27	35.00	22.97	35.36	27.45	40.81	24.55	46.63	17.61	42.17	20.69	-	-
5/26/1999	-	-	-	-	-	-	31.95	25.93	34.00	23.84	35.50	22.47	35.80	27.01	41.06	24.30	43.06	21.18	41.60	21.26	57.84	5.06
6/1/1999	-	-	-	-	-	-	32.78	25.10	34.80	23.04	36.08	21.89	36.40	26.41	41.66	23.70	43.70	20.54	42.20	20.66	-	-
6/15/1999	32.80	25.38	-	-	-	-	35.57	22.31	36.18	21.66	37.77	20.20	37.70	25.11	43.00	22.36	49.26	14.98	44.60	18.26	-	-
6/22/1999	32.10	26.08	33.00	25.41	38.00	20.47	-	-	-	-	-	-	38.10	24.71	43.43	21.93	49.50	14.74	45.04	17.82	-	-

Abbreviations

DTW= depth to water from MPE (ft)

Elev.= elevation relative to Mean Sea Level (ft)

MPE= measuring point elevation relative to Mean Sea Level (ft)

Notes:

* The MPE for PT 8 was 1.65 in. higher than the point measured by GPS at 68.890 ft.

† MPE for 20J04-20J06 and 20R03-20R05 were measured by Union Pacific Railroad (or consultant).

‡ The MPE for Mowry 8 was 2 in. higher than the point measured by GPS at 64.075ft.

APPENDIX B
TABLES OF WATER QUALITY

Table B-1
Field measurements

Sample I.D.	Date	Dissolved Oxygen (mg/L)	Temp (C)	Conductivity (µmhos/cm)	Redox* (mV)	Color/Clarity	pH
SURFACE WATER							
Bunting East at 0'	12/16/1997	-	11.4	724	-	-	8.50
Bunting East at 3.3'	12/16/1997	-	11.2	721	-	-	8.50
Bunting East at 6.6'	12/16/1997	-	11.4	690	-	-	8.50
Bunting East at 9.8'	12/16/1997	-	10.8	725	-	-	8.50
Bunting East at 13.1'	12/16/1997	-	10.9	709	-	-	8.50
Bunting East at 29.5'	12/16/1997	-	10.8	722	-	-	8.50
Bunting- West Mid at 3.3	12/30/1997	11.1	8.92	767	368		7.92
Bunting- West Mid at 16.5'	12/30/1997	10.9	8.89	767	367		7.93
Bunting- West Mid at 33'	12/30/1997	10.7	8.86	767	367		7.92
Bunting -S.W. at 3.3'	12/30/1997	11.1	8.91	767	392		7.94
Bunting -S.W. at 16.5'	12/30/1997	11.0	8.86	768	395		7.95
Bunting -S.W. at 26.4'	12/30/1997	10.8	8.84	768	395		7.95
Kaiser AHF- Middle at 0'	12/16/1997	-	12.5	699	-	-	8.40
Kaiser AHF- Middle at 3.3'	12/16/1997	-	12.6	680	-	-	8.30
Kaiser AHF- Middle at 9.8'	12/16/1997	-	12.3	706	-	-	8.40
Kaiser AHF- Middle at 29.5'	12/16/1997	-	12.3	706	-	-	8.40
Kaiser AHF- Middle at 52.5'	12/16/1997	-	12.0	707	-	-	8.40
Kaiser AHF - Middle at 3.3'	12/30/1997	9.3	10.8	760	425		7.88
Kaiser AHF - Middle at 16.5'	12/30/1997	9.3	10.5	760	425		7.85
Kaiser AHF - Middle at 33'	12/30/1997	9.2	10.5	760	424		7.84
Kaiser AHF - Middle at 59.4'	12/30/1997	9.1	10.3	760	424		7.84
Kaiser AHF- South at 3.3'	12/30/1997	9.5	10.9	760	443		7.85
Kaiser AHF- South at 16.5'	12/30/1997	9.4	10.5	760	441		7.85
Kaiser AHF- South at 33'	12/30/1997	9.2	10.4	760	438		7.83
Kaiser AHF- South at 42.9'	12/30/1997	9.2	10.3	761	440		7.83
Pit S at 1'	8/17/1998	8.8	24.0	751	331.4		8.92
Pit S at 2'	8/17/1998	8.8	24.0	751	333		8.91
Pit S at 3'	8/17/1998	8.8	24.0	751	333.6		8.92
Pit S at 6'	8/17/1998	8.7	24.0	750	334		8.92
Pit S at 15'	8/17/1998	8.6	23.9	751	334.6		8.91
Pit S at 20'	8/17/1998	1.7	22.9	749	342.5		8.27
Pit S at 23'	8/17/1998	0.2	22.4	746	338.9		8.10
Pit S at 30'	8/17/1998	0.2	19.4	703	143.8		7.84
Pit S -N.E. at 2'	8/20/1998	7.7	23.1	753	310.7		8.81
Pit S -N.E. at 3'	8/20/1998	7.6	23.1	753	309.9		8.81
Pit S -N.E. at 15'	8/20/1998	7.1	23.0	752	309.9		8.78
Pit S - N.E. at 20'	8/20/1998	5.7	22.9	751	306.8		8.67

(continued)

Table B-1 (continued)

Sample I.D.	Date	Dissolved Oxygen (mg/L)	Temp (C)	Conductivity (µmhos/cm)	Redox* (mV)	Color/Clarity	pH
Pit S. -N.E. at 25'	8/20/1998	1.5	24.9	740	315.6		8.15
Pit S - N.E. at 30'	8/20/1998	0.2	19.0	704	105.1		7.79
Pit S - South at 1'	8/20/1998	7.4	23.2	750	268.9		8.81
Pit S - South at 10'	8/20/1998	7.5	23.2	750	270.1		8.81
Pit S - South at 30'	8/20/1998	0.2	19.0	705	94.2		7.84
Pit S - South at 40'	8/20/1998	0.2	16.3	672	74.9		7.77
Pit S- South at 48'	8/20/1998	0.2	14.6	652	57.8		7.62
MONITORING WELLS							
20J04	8/24/1998	0.7	21.8	852	-	-	7.46
20J04	9/1/1998	0.3	21.8	909	-	-	7.41
20J04	9/15/1998	0.4	21.5	870	247.8	clear	7.49
20J04	9/29/1998	0.0	22.9	881	-	clear	7.27
20J04	10/13/1998	0.0	23.6	857	-	slight yellow	7.54
20J04	10/20/1998	0.5	23.2	869	-	clear	7.23
20J04	10/28/1998	0.5	22.9	941	-	slight yellow	7.21
20J04	11/9/1998	0.2	21.7	860	323.1	slight yellow	7.22
20J04	12/1/1998	0.1	20.3	821	-	slight yellow	7.33
20J04	1/5/1999	0.55	16.5	771	-	-	7.42
20J06	8/24/1998	0.7	20.1	774	-	-	7.63
20J06	9/22/1998	0.2	20.8	798	-	-	7.67
20J06	10/6/1998	0.0	19.4	843	-	clear	7.77
20J06	11/17/1998	1.7	18.2	824	298.8	clear	7.32
20J06	1/5/1999	0.4	19.3	813	-	-	7.44
20R03	8/24/1998	0.6	25.2	796	-	-	7.45
20R03	9/1/1998	0.5	25.6	840	-	slight yellow	7.48
20R03	9/15/1998	1.0	25.0	883	238.5	clear	7.50
20R03	10/2/1998	0.2	23.6	813	-	yellowish	7.31
20R03	10/13/1998	0.8	24.3	793	-	-	7.68
20R03	10/20/1998	0.1	25.2	837	-	clear	7.29
20R03	10/28/1998	1.7	23.4	912	-	-	7.27
20R03	11/9/1998	0.1	21.3	768	293.2	slight yellow	7.42
20R03	12/1/1998	0.2	19.7	781	-	clear	7.48
20R03	1/5/1999	2.25	18.7	746	-	light yellow	7.53
20R04	9/22/1998	0.3	9.8	679	-	-	7.48
20R04	10/6/1998	0.0	20.0	715	-	-	7.70
20R04	11/17/1998	1.2	19.0	739	316.0	clear	7.47
20R04	1/5/1999	0.55	19.5	745	-	-	7.40
20R07	9/15/1998	0.6	20.3	759	-	-	7.33
20R07	9/29/1998	1.3	18.6	832	-	-	7.32
20R07	10/13/1998	0.2	19.4	823	-	-	7.55
20R07	11/3/1998	0.5	19.2	824	-	-	7.25
20R07	11/17/1998	1.3	18.8	837	304.2	-	7.13
20R07	11/23/1998	0.5	17.8	850	-	clear	7.17

(continued)

Table B-1 (continued)

Sample I.D.	Date	Dissolved Oxygen (mg/L)	Temp (C)	Conductivity (µmhos/cm)	Redox* (mV)	Color/Clarity	pH
20R07	1/27/1999	0.7	18.7	804	-	clear	7.10
20R07	4/6/1999	0.8	18.5	791	-	-	7.09
20R07	6/1/1999	0.7	19.5	744	-	-	7.39
20R07	10/20/1999	0.4	18.2	710	-	clear	6.75
21H11	8/24/1998	0.6	21.1	1529	-	-	6.94
21H11	9/1/1998	0.9	21.3	1683	-	-	6.88
21H11	9/15/1998	0.8	21.8	1380	221.8	clear	6.99
21H11	9/22/1998	0.5	19.5	1480	-	-	6.93
21H11	9/29/1998	0.0	19.2	1494	-	clear	6.86
21H11	10/6/1998	0.0	20.2	1547	-	-	7.04
21H11	11/9/1998	0.2	19.7	1515	-	clear	6.86
21J03	12/16/1997	0.5	17.5	670	327.0	clear	7.40
21J03	12/23/1997	1.6	16.4	-	-	-	-
21J03	12/30/1997	0.9	17.2	-	-	-	-
21J03	1/6/1998	0.1	17.4	798	-	clear	7.50
21J03	1/13/1998	1.1	17.9	986	-	-	7.50
21J03	1/20/1998	2.1	17.2	1066	-	-	7.60
21J03	2/17/1998	1.5	17.2	880	-	-	7.65
21J03	2/24/1998	3.0	17.1	876	-	-	7.50
21J03	3/3/1998	1.9	19.1	855	-	clear	7.50
21J03	3/17/1998	3.0	24.1	740	-	-	7.60
21J03	3/31/1998	4.0	16.5	704	-	-	7.60
21J03	4/7/1998	2.6	20.0	860	-	clear	7.70
21J03	4/21/1998	3.4	20.7	839	-	-	7.70
21J03	5/26/1998	0.8	18.5	709	-	-	7.50
21J03	9/1/1998	0.5	20.6	871	-	-	7.35
21J03	9/15/1998	0.2	19.1	813	-	-	7.53
21J03	9/22/1998	0.6	19.5	817	-	-	7.39
21J03	10/13/1998	0.3	18.9	838	-	clear	7.59
21J03	10/20/1998	0.6	19.5	834	-	-	7.37
21J03	12/8/1998	0.0	16.8	857	-	clear	7.38
21J03	12/29/1998	1.0	17.5	849	-	clear	7.18
21J03	1/27/1999	0.9	18.0	835	-	clear	7.19
21J27	2/24/1998	3.2	19.6	860	-	-	7.40
21J27	3/3/1998	4.4	21.4	823	-	clear	7.50
21J27	3/24/1998	3.8	23.0	803	-	clear	7.70
21J27	3/31/1998	5.0	18.3	919	-	-	7.40
21J27	4/7/1998	5.4	19.6	849	-	-	7.60
21J27	4/21/1998	3.1	21.5	821	-	-	7.60
21J27	4/28/1998	4.0	20.0	848	-	clear	-
21J27	5/5/1998	3.8	20.4	822	-	-	7.50
21J27	5/19/1998	4.0	18.5	717	-	-	7.32
21J27	6/2/1998	4.6	19.1	664	-	-	7.50
21J27	6/16/1998	3.4	24.4	641	-	-	7.27
21J27	7/7/1998	6.0	19.5	797	-	-	7.24

(continued)

Table B-1 (continued)

Sample I.D.	Date	Dissolved Oxygen (mg/L)	Temp (C)	Conductivity (µmhos/cm)	Redox* (mV)	Color/Clarity	pH
21J27	7/28/1998	2.3	19.4	835	-	-	7.22
21J27	10/6/1998	1.2	21.5	890	-	-	7.45
21J27	11/3/1998	1.7	22.8	881	-	-	7.37
21J27	12/1/1998	1.3	21.4	918	-	clear	7.41
21J27	3/10/1999	-	19.2	862	-	-	7.00
21J27	6/1/1999	3.8	17.1	760	-	-	7.40
21K05	1/14/1998	-	-	-	-	very turbid	-
21K05	1/20/1998	0.7	19.3	980	-	mod'ly turbid	7.59
21K05	1/27/1998	0.4	21.3	890	-	clear	7.38
21K05	2/4/1998	1.4	19.1	880	-	-	7.52
21K05	2/10/1998	2.1	20.6	960	-	-	7.50
21K05	2/18/1998	-	19.5	836	-	-	7.60
21K05	2/24/1998	2.2	20.6	790	-	clear	7.50
21K05	3/10/1998	3.1	23.9	676	-	clear	7.70
21K05	3/24/1998	1.9	21.6	780	-	clear	7.60
21K05	4/8/1998	1.6	22.5	712	-	-	7.70
21K05	4/21/1998	2.3	23.9	710	-	-	7.70
21K05	4/28/1998	2.6	22.6	714	-	clear	-
21K05	5/5/1998	1.4	21.4	712	-	clear	7.60
21K05	5/26/1998	1.1	19.8	615	-	-	7.60
21K05	7/14/1998	0.8	20.9	892	-	-	7.44
21K05	9/22/1998	1.0	20.1	819	-	-	7.35
21K05	10/13/1998	0.0	19.7	865	-	clear	7.75
21K05	11/3/1998	0.3	19.8	877	-	-	7.49
21K05	11/17/1998	1.0	19.4	873	286.2	slight yellow	7.24
21K05	12/29/1998	0.8	18.9	862	-	-	7.29
21K05	1/27/1999	0.8	20.1	844	-	clear	7.01
21L03	12/23/1997	2.8	18.2	-	-	-	-
21L03	12/30/1997	1.0	19.8	791	-	-	-
21L03	1/6/1998	0.6	18.2	825	-	-	7.30
21L03	1/27/1998	0.4	20.8	887	-	clear	7.30
21L03	2/4/1998	0.2	18.1	887	348.0	clear	7.40
21L03	2/10/1998	4.0	-	140	-	-	7.62
21L03	2/17/1998	0.8	18.5	886	-	clear	7.50
21L03	2/24/1998	1.3	20.0	890	-	-	7.40
21L03	3/3/1998	0.6	19.0	883	-	clear	7.40
21L03	3/10/1998	3.1	21.0	839	-	-	7.50
21L03	3/17/1998	4.0	18.9	890	-	-	7.50
21L03	3/24/1998	4.1	18.5	854	-	-	7.60
21L03	3/31/1998	2.2	18.2	890	-	-	7.50
21L03	5/12/1998	0.9	17.8	840	-	-	7.00
21L03	6/30/1998	1.8	22.5	877	-	-	7.30
21L03	7/28/1998	0.0	21.9	808	-	-	7.28
21L03	10/28/1998	0.8	18.5	905	-	-	7.51
21L03	1/12/1999	1.1	18.2	828	-	-	7.64

(continued)

Table B-1 (continued)

Sample I.D.	Date	Dissolved Oxygen (mg/L)	Temp (C)	Conductivity (µmhos/cm)	Redox* (mV)	Color/Clarity	pH
21L05	12/2/1997	2.1	16.1	10.5	342.7	clear	7.60
21L05	12/23/1997	2.8	18.2	-	-	-	-
21L05	1/13/1998	1.3	18.3	969	-	-	7.50
21L05	1/20/1998	0.7	17.1	1005	-	-	7.50
21L05	1/27/1998	0.8	19.4	928	-	clear	7.40
21L05	2/4/1998	0.5	19.3	869	315.1	clear	7.53
21L05	2/10/1998	2.3	18.1	875	-	clear	7.50
21L05	2/17/1998	2.0	18.5	925	-	clear	7.55
21L05	2/25/1998	1.9	19.4	874	-	clear	7.45
21L05	3/10/1998	1.5	21.3	855	-	-	7.50
21L05	3/24/1998	2.8	21.5	856	-	-	7.50
21L05	4/8/1998	2.5	20.5	862	-	-	7.60
21L05	4/21/1998	2.1	19.4	860	-	-	7.70
21L05	4/28/1998	2.4	21.6	826	-	clear	-
21L05	5/12/1998	1.8	19.5	815	-	-	7.50
21L05	5/26/1998	0.6	18.2	689	-	-	7.50
21L05	6/30/1998	0.8	20.2	782	-	-	7.30
21L05	7/14/1998	1.0	18.8	810	-	-	7.63
21L05	10/29/1998	0.3	18.1	856	-	-	7.48
21l05	3/16/1999	0.3	19.7	889	-	-	7.17
21L06	12/2/1997	1.8	21.6	11.2	349.3	slightly turbid	7.45
21L06	12/30/1997	0.5	20.0	805	-	-	-
21L06	1/6/1998	0.4	20.1	802	-	clear	7.40
21L06	1/13/1998	0.5	19.5	998	-	yellowish	7.39
21L06	1/20/1998	1.0	19.0	1006	-	-	7.40
21L06	1/27/1998	1.2	21.3	864	-	-	7.36
21L06	2/4/1998	1.4	19.0	930	410.0	clear	7.40
21L06	2/10/1998	1.3	18.3	870	-	-	7.40
21L06	2/17/1998	2.0	19.1	860	-	clear	7.50
21L06	2/25/1998	0.8	20.6	830	-	clear	7.40
21L06	3/10/1998	3.1	22.9	800	-	-	7.60
21L06	3/24/1998	1.7	22.8	797	-	clear	7.40
21L06	4/8/1998	2.1	19.6	808	-	-	7.40
21L06	4/21/1998	2.4	20.3	725	-	-	7.50
21L06	5/12/1998	1.6	20.1	760	-	-	7.50
21L06	5/26/1998	0.6	20.3	609	-	-	7.40
21L06	6/30/1998	0.6	21.2	719	-	-	7.40
21L06	7/14/1998	1.1	20.2	738	-	-	7.55
21L06	11/9/1998	0.1	19.9	853	335.6	-	7.18
21L07	12/2/1997	2.1	22.2	10.4	329.0	-	7.48
21L07	12/30/1997	0.8	20.3	777	-	-	-
21L07	1/13/1998	1.3	18.8	981	373.9	-	7.52
21L07	1/20/1998	0.3	19.0	751	-	slightly turbid	7.50
21L07	1/27/1998	0.3	20.3	915	-	-	7.34
21L07	2/4/1998	0.6	18.7	856	337.0	clear	7.50

(continued)

Table B-1 (continued)

Sample I.D.	Date	Dissolved Oxygen (mg/L)	Temp (C)	Conductivity (µmhos/cm)	Redox* (mV)	Color/Clarity	pH
21L07	2/10/1998	2.7	17.9	845	-	clear	7.50
21L07	2/17/1998	2.2	20.1	836	-	clear	7.60
21L07	2/25/1998	0.7	21.2	798	-	clear	7.40
21L07	3/10/1998	2.1	22.0	785	-	-	7.50
21L07	3/17/1998	0.5	19.7	794	-	-	7.50
21L07	3/24/1998	2.0	21.0	808	-	clear	7.50
21L07	4/8/1998	1.4	19.6	750	-	-	7.40
21L07	10/29/1998	0.2	20.8	921	-	-	7.29
21L08	12/16/1997	3.5	15.2	777	310.0	-	6.80
21L08	12/23/1997	5.0	14.7	-	-	-	-
21L08	12/30/1997	4.0	14.0	741	-	-	-
21L08	1/6/1998	4.6	13.7	707	-	-	7.86
21L08	1/13/1998	4.1	13.8	923	377.0	-	7.88
21L08	1/20/1998	0.6	18.6	960	-	-	7.70
21L08	1/27/1998	2.1	17.1	831	-	-	7.56
21L08	2/10/1998	1.9	14.8	843	-	-	7.52
21L08	2/17/1998	2.0	17.7	835	-	clear	7.80
21L08	2/25/1998	1.5	17.7	848	-	clear	7.60
21L08	3/10/1998	6.0	17.9	650	-	-	7.90
21L08	3/17/1998	5.3	18.1	698	-	-	7.90
21L08	3/24/1998	3.7	18.4	792	-	clear	7.70
21L08	4/8/1998	2.6	17.3	871	-	-	7.70
21L08	4/21/1998	3.6	19.3	770	-	-	7.70
21L08	5/12/1998	0.6	17.5	863	-	-	7.50
27D08	1/6/1998	2.1	17.4	-	-	-	7.20
27D08	4/7/1998	5.0	19.4	1218	-	clear	7.36
27D08	6/2/1998	3.5	21.2	1043	-	-	7.30
27D08	6/9/1998	3.1	23.2	1028	-	-	7.08
27D08	7/7/1998	3.5	21.9	1204	-	-	7.08
27D08	7/14/1998	3.5	21.9	1204	-	-	7.08
28D08	7/20/1999	0.1	19.1	761	-	slightly turbid	7.20
28D11	11/23/1998	-	17.6	766	-	clear	7.45
28D11	12/8/1998	0.0	17.6	754	-	-	7.48
28D11	1/5/1999	0.6	17.4	768	-	-	7.48
28D11	3/10/1999	0	17.8	755	-	-	7.40
28D11	4/20/1999	0.0	18.6	735	-	clear	7.18
28D11	5/25/1999	0.1	19.5	779	-	clear	7.27
28D11	7/6/1999	0.0	19.0	770	-	clear	7.28
28D12	11/24/1998	-	18.8	845	-	clear	7.38
28D12	12/8/1998	0.0	17.9	-	-	-	7.43
28D12	12/29/1998	1.1	18.4	845	-	clear	7.06
28D12	2/24/1999	0.1	17.9	829	-	clear	7.27
28D12	4/6/1999	0.4	18.0	837	-	-	7.30
28D12	5/25/1999	0.15	19.9	826	-	clear	7.16
28D12	7/6/1999	0.0	19.2	814	-	clear	7.11

(continued)

Table B-1 (continued)

Sample I.D.	Date	Dissolved Oxygen (mg/L)	Temp (C)	Conductivity (µmhos/cm)	Redox* (mV)	Color/Clarity	pH
28D12	10/20/1999	0.3	18.5	830	-	clear	6.77
PRODUCTION WELLS							
MOWRY 1	8/4/1998	1.8	18.8	916	-	-	7.41
MOWRY 1	8/18/1998	1.5	17.7	847	-	-	7.54
MOWRY 1	10/14/1998	0.9	17.4	851	-	-	7.54
MOWRY 1	10/29/1998	1.3	17.2	904	-	-	7.30
MOWRY 1	11/9/1998	1.6	17.1	854	-	-	7.28
MOWRY 1	12/9/1998	1.5	16.7	-	-	-	7.30
MOWRY 1	12/22/1998	2.3	16.5	851	-	-	7.21
MOWRY 1	1/12/1999	2.7	16.8	842	-	-	7.25
MOWRY 1	1/27/1999	2.2	17.0	838	-	clear	6.97
MOWRY 1	2/24/1999	1.2	17.4	874	-	-	7.30
MOWRY 1	4/6/1999	1.1	16.9	858	-	-	7.38
MOWRY 1	4/22/1999	1.4	17.4	830	-	clear	7.20
MOWRY 1	6/8/1999	0.9	17.9	840	-	clear	7.25
MOWRY 1	7/20/1999	1.4	17.9	842	-	-	7.31
MOWRY 1	12/8/1999	2.5	15.7	862	-	-	6.89
MOWRY 3	8/18/1998	1.3	18.2	963	-	-	7.24
MOWRY 6	12/16/1998	2.7	17.6	1305	-	-	7.26
MOWRY 6	6/8/1999	1.2	17.9	1415	-	mod'ly turbid	7.15
MOWRY 6	7/20/1999	1.3	17.6	1199	-	-	7.21
MOWRY 7	5/19/1998	3.7	20.3	801	-	-	-
MOWRY 7	8/4/1998	2.1	19.0	824	-	-	7.24
MOWRY 7	10/14/1998	1.6	17.7	899	-	slightly turbid	7.45
MOWRY 7	10/29/1998	5.4	25.0	961	-	very turbid	7.30
MOWRY 7	11/9/1998	2.4	17.3	894	-	-	7.29
MOWRY 7	12/9/1998	2.1	17.2	-	-	-	7.38
MOWRY 7	1/12/1999	2.7	16.5	879	-	clear	7.35
MOWRY 7	2/2/1999	-	16.9	926	-	-	7.17
MOWRY 7	2/24/1999	2.1	17.7	905	-	-	7.32
MOWRY 7	4/6/1999	2.2	16.8	901	-	-	7.29
MOWRY 7	4/22/1999	2.7	17.5	894	-	clear	7.07
MOWRY 7	6/15/1999	2.0	17.2	879	-	slightly turbid	7.29
MOWRY 7	12/8/1999	2.2	16.0	882	-	-	6.58
MOWRY 9	12/16/1998	1.7	17.2	929	-	clear	7.33
MOWRY 9	6/8/1999	0.4	17.5	890	-	clear	7.19
MOWRY 9	7/20/1999	0.3	17.0	825	-	-	7.14
PT-1	2/17/1998	2.5	18.5	988	-	-	7.60
PT-1	4/28/1998	4.1	20.9	1069	-	clear	-
PT-1	5/26/1998	3.0	20.3	880	-	-	7.30
PT-1	6/2/1998	2.0	20.0	825	-	-	7.30
PT-1	6/9/1998	1.9	20.9	821	-	turbid	7.36
PT-1	6/23/1998	3.1	19.6	1060	-	-	7.27
PT-1	6/30/1998	3.2	18.6	1081	-	-	7.60
PT-1	7/7/1998	4.5	21.2	1069	-	-	7.21

(continued)

Table B-1 (continued)

Sample I.D.	Date	Dissolved Oxygen (mg/L)	Temp (C)	Conductivity (µmhos/cm)	Redox* (mV)	Color/Clarity	pH
PT-1	7/21/1998	4.4	19.9	1108	-	slightly turbid	7.22
PT-1	8/25/1998	3.4	19.1	1232	-	-	7.21
PT-1	10/2/1998	2.6	19.2	1149	-	-	7.32
PT-1	12/1/1998	1.5	18.2	966	-	clear	7.35
PT-1	12/22/1998	1.5	17.6	912	-	-	7.23
PT-1	3/9/1999	2.2	18.5	1090	-	slightly turbid	7.14
PT-1	12/15/1999	3.0	16.6	1080	-	-	6.89
PT-2	4/28/1998	3.2	22.5	927	-	clear	-
PT-2	5/5/1998	3.6	19.3	945	-	-	7.80
PT-2	2/2/1999	-	17.9	911	-	-	7.38
PT-2	3/16/1999	0.8	18.2	902	-	-	7.22
PT-2	4/27/1999	1.5	18.4	968	-	slightly turbid	7.15
PT-2	12/8/1999	0.7	16.4	846	-	-	6.86
PT-3	2/17/1998	0.9	19.0	900	-	-	7.55
PT-3	2/24/1998	3.4	18.3	937	-	-	6.92
PT-3	3/3/1998	2.6	17.6	923	-	clear	7.79
PT-3	3/17/1998	4.2	18.2	926	-	-	8.00
PT-3	3/24/1998	1.8	22.2	880	-	clear	7.50
PT-3	3/31/1998	2.8	17.3	915	-	-	7.70
PT-3	4/7/1998	-	18.6	-	222.0	-	7.70
PT-3	4/15/1998	3.4	18.6	949	-	-	7.90
PT-3	4/21/1998	2.7	20.8	468	-	-	7.70
PT-3	4/28/1998	2.0	20.4	909	-	clear	-
PT-3	5/5/1998	2.2	19.3	870	-	-	7.80
PT-3	5/12/1998	2.6	18.3	917	-	-	7.90
PT-3	5/19/1998	1.5	19.5	773	-	-	7.42
PT-3	5/26/1998	2.1	18.6	742	-	-	8.00
PT-3	6/2/1998	0.6	20.9	714	-	-	7.50
PT-3	6/9/1998	1.2	21.0	695	-	-	7.42
PT-3	6/16/1998	1.4	20.9	715	-	clear	7.47
PT-3	6/23/1998	1.5	19.6	912	-	-	7.30
PT-3	6/30/1998	2.2	18.4	928	-	-	7.30
PT-3	7/7/1998	2.3	21.5	958	-	-	7.25
PT-3	7/14/1998	1.7	21.8	897	-	-	7.40
PT-3	7/28/1998	1.6	20.6	1001	-	-	7.33
PT-3	8/4/1998	2.1	20.0	972	-	-	7.38
PT-3	8/25/1998	2.3	19.1	1008	-	-	7.28
PT-3	10/2/1998	1.4	18.7	970	-	-	7.19
PT-3	11/3/1998	1.1	18.6	909	-	clear	7.39
PT-3	12/1/1998	0.2	18.3	859	-	clear	7.23
PT-3	12/16/1998	0.7	18.8	853	-		7.29
PT-3	2/2/1999	-	18.2	897	-	-	7.14
PT-3	3/9/1999	0.1	18.6	874	-	-	7.19
PT-3	12/15/1999	0.5	16.6	795	-	-	6.75
PT-4	7/7/1998	2.6	19.8	966	-	-	7.27

(continued)

Table B-1 (continued)

Sample I.D.	Date	Dissolved Oxygen (mg/L)	Temp (C)	Conductivity (µmhos/cm)	Redox* (mV)	Color/Clarity	pH
PT-4	7/21/1998	2.2	19.7	986	-	slightly turbid	7.29
PT-4	8/11/1998	3.1	19.8	1149	-	-	7.37
PT-4	8/18/1998	2.6	19.0	1041	-	-	7.38
PT-4	9/2/1998	2.5	19.4	1145	-	-	7.29
PT-4	10/21/1998	1.3	19.9	1008	-	-	7.12
PT-4	11/23/1998	1.0	17.8	968	-	clear	7.21
PT-4	12/22/1998	0.9	18.4	856	-	-	7.35
PT-4	4/6/1999	1.6	18.6	992	-	-	7.35
PT-4	4/27/1999	1.5	19.0	966	-	slightly turbid	7.10
PT-4	12/8/1999	1.5	17.0	922	-	-	6.79
PT-5	6/30/1998	1.8	18.4	838	-	-	7.50
PT-5	7/7/1998	1.8	19.8	858	-	-	7.59
PT-5	7/21/1998	1.9	18.9	851	-	-	7.82
PT-5	8/4/1998	1.5	19.7	799	-	-	7.36
PT-5	8/18/1998	1.1	18.7	833	-	-	7.38
PT-5	9/2/1998	1.6	18.9	994	-	-	7.42
PT-5	10/2/1998	1.3	18.2	856	-	-	7.06
PT-5	10/21/1998	0.9	18.3	839	-	-	7.32
PT-5	11/4/1998	1.0	18.2	851	-	clear	7.31
PT-5	12/1/1998	0.8	18.5	849	-	clear	7.29
PT-5	12/16/1998	1.1	19.1	850	-	clear	7.26
PT-5	2/2/1999	-	18.7	906	-	-	7.26
PT-5	4/6/1999	1.1	17.9	895	-	-	7.28
PT-5	4/27/1999	0.9	18.7	863	-	slightly turbid	7.21
PT-5	6/1/1999	0.9	19.0	805	-	-	7.16
PT-6	6/9/1998	0.6	19.8	649	-	-	7.64
PT-6	6/16/1998	1.4	22.3	677	-	-	7.28
PT-6	6/23/1998	1.6	18.5	841	-	-	7.60
PT-6	6/30/1998	1.3	17.9	814	-	-	7.30
PT-6	7/7/1998	1.4	19.7	772	-	-	7.30
PT-6	7/21/1998	1.0	18.3	861	-	-	7.26
PT-6	8/18/1998	1.1	18.3	797	-	-	7.26
PT-6	9/2/1998	0.8	18.4	857	-	-	7.50
PT-6	10/2/1998	1.0	18.2	846	-	-	7.18
PT-6	11/3/1998	0.3	18.4	839	-	clear	7.31
PT-6	11/24/1998	-	19.0	1022	-	clear	7.40
PT-6	12/22/1998	0.8	18.6	840	-	-	7.32
PT-6	12/8/1999	0.8	17.3	805	-	-	6.77
PT-7	2/17/1998	1.7	20.0	888	-	-	7.80
PT-7	2/24/1998	3.0	18.1	795	-	-	7.40
PT-7	3/3/1998	2.2	20.1	862	-	clear	7.60
PT-7	3/10/1998	2.9	20.0	852	-	-	7.40
PT-7	3/24/1998	3.8	22.1	870	-	clear	7.50
PT-7	3/31/1998	3.0	17.4	883	-	-	7.60
PT-7	4/7/1998	-	18.7	-	223.0	-	7.60

(continued)

Table B-1 (continued)

Sample I.D.	Date	Dissolved Oxygen (mg/L)	Temp (C)	Conductivity (μmhos/cm)	Redox* (mV)	Color/Clarity	pH
PT-7	4/15/1998	2.1	19.0	863	-	-	7.70
PT-7	4/21/1998	1.2	23.6	819	-	-	7.60
PT-7	5/19/1998	0.9	19.9	719	-	-	7.26
PT-7	6/9/1998	0.8	23.9	682	-	-	7.38
PT-7	6/16/1998	0.9	24.5	686	-	-	7.29
PT-7	6/30/1998	1.1	18.2	786	-	-	7.40
PT-7	7/7/1998	1.7	20.4	780	-	-	7.29
PT-7	7/21/1998	1.4	18.5	776	-	-	7.31
PT-7	8/4/1998	0.8	20.2	783	-	-	7.20
PT-7	8/18/1998	0.9	19.2	771	-	-	7.21
PT-7	9/2/1998	0.3	18.9	842	-	-	7.45
PT-7	10/21/1998	0.5	19.0	823	-	-	7.19
PT-7	11/4/1998	0.9	19.4	829	-	clear	7.34
PT-7	12/22/1998	1.1	18.7	854	-	-	7.43
PT-7	1/12/1999	1.3	18.1	862	-	-	7.30
PT-7	2/2/1999	-	18.3	899	-	-	7.16
PT-7	3/9/1999	0.3	19.1	867	-	-	7.17
PT-7	4/27/1999	0.6	19.0	830	-	clear	7.18
PT-7	6/1/1999	0.6	18.1	811	-	-	7.39
PT-8	7/14/1998	1.2	20.3	844	-	-	7.48
PT-8	7/21/1998	1.1	18.3	819	-	-	7.32
PT-8	8/18/1998	1.1	18.5	788	-	-	7.30
PT-8	8/25/1998	1.5	18.4	774	-	-	7.66
PT-8	3/9/1999	0	18.5	854	-	-	7.21
PT-8	6/1/1999	0.4	18.6	831	-	-	7.29
PT-8	12/8/1999	0.6	16.5	815	-	-	7.10

*values adjusted in order to be relative to an H_2/H^+ standard electrode

Table B-2

General water quality parameters

Sample I.D.	Date Sampled	Alkalinity and Hardness (mg/L as $CaCO_3$)				
		Alkalinity				Total Hardness
		Total	Bicarb.	Carbonate	Hydrox. (as OH)	
Pit S at 5'*	4/16/1997	-	150	16	<0.1	200
Pit S at 22'*	4/16/1997	-	170	<2	<0.1	210
Pit S at 52'*	4/16/1997	-	180	<2.0	<0.1	210
20J04	5/6/1997	-	-	-	-	-
20J06	5/6/1997	-	-	-	-	-
20J05	5/8/1997	-	-	-	-	-
20J05†	5/8/1997	-	-	-	-	-
20R03	5/8/1997	-	-	-	-	-
20R03†	5/8/1997	-	-	-	-	-
20R04	5/8/1997	-	-	-	-	-
20R04†	5/8/1997	-	-	-	-	-
20R05	5/8/1997	-	-	-	-	-
20R05†	5/8/1997	-	-	-	-	-
20J04	6/18/1997	-	-	-	-	-
20J05	6/18/1997	-	-	-	-	-
20J06	6/18/1997	-	-	-	-	-
20R04	6/18/1997	-	-	-	-	-
20R05	6/18/1997	-	-	-	-	-
Bunting South at 3.3'	12/29/1997	-	210	<2	<0.1	580
Bunting South at 3.3'†	12/29/1997	-	-	-	-	270
Bunting South at 16.4'	12/29/1997	-	210	<2	<0.1	600
Bunting South at 16.4'†	12/29/1997	-	-	-	-	280
Bunting Middle at 3.3'	12/29/1997	-	220	<2	<0.1	580
Bunting Middle at 3.3'†	12/29/1997	-	-	-	-	270
Bunting Middle at 32.8'	12/29/1997	-	260	<2	<0.1	700
Bunting Middle at 32.8'†	12/29/1997	-	-	-	-	280
Kaiser AHF South at 3.3'	12/29/1997	-	180	<2	<0.1	700
Kaiser AHF South at 3.3'†	12/29/1997	-	-	-	-	270
Kaiser AHF South at 36.1'	12/29/1997	-	220	<2	<0.1	1000
Kaiser AHF South at 36.1'†	12/29/1997	-	-	-	-	260
Kaiser AHF Middle at 3.3'	12/29/1997	-	220	<2	<0.1	640
Kaiser AHF Middle at 3.3'†	12/29/1997	-	-	-	-	270
Kaiser AHF Middle at 62.3'	12/29/1997	-	220	<2	<0.1	1200
Kaiser AHF Middle at 62.3'†	12/29/1997	-	-	-	-	260
21L07	1/13/1998	260	260	<2	<0.1	210
21L08	1/13/1998	-	230	<2	<0.1	240
21L05	2/4/1998	-	250	<2	<0.1	260
21L05‡	2/4/1998	-	-	-	-	-
21L06	2/4/1998	-	260	<2	<0.1	250
21L06‡	2/4/1998	-	-	-	-	-
21L07	2/4/1998	-	260	<2	<0.1	250
21L07‡	2/4/1998	-	-	-	-	-

(continued)

Table B-2 (continued)

Sample I.D.	Date Sampled	Alkalinity and Hardness (mg/L as $CaCO_3$)				
		Alkalinity				Total Hardness
		Total	Bicarb.	Carbonate	Hydrox. (as OH)	
21K05	2/4/1998	-	240	<2	<0.1	240
21K05‡	2/4/1998	-	-	-	-	-
21L03	3/3/1998	-	220	<2	<0.1	260
21L03†	3/3/1998	-	-	-	-	-
21J03	3/3/1998	-	210	<2	<0.1	250
21J03†	3/3/1998	-	-	-	-	-
PT-7	4/7/1998	-	220	<4	<0.2	260
PT-3	4/7/1998	-	230	<4	<0.2	290
PT-1	6/9/1998	740	740	<2	<0.1	330
PT-3	6/9/1998	330	330	<2	<0.1	260
PT-6	6/9/1998	240	240	<2	<0.1	240
PIT S Northeast at 10'	8/25/1998	180	160	16	<5	160
PIT S Northeast at 40'	8/25/1998	190	<5	<5	<5	180
RD-3 Upstream-750' at 2'	8/25/1998	230	220	16	<5	210
RD-3 Upstream-Immediate	8/25/1998	230	230	<5	<5	210
20J04	9/15/1998	210	210	<5	<5	260
20R03	9/15/1998	110	86	24	<5	240
21H11	9/15/1998	740	740	<5	<5	<1
20J06	11/17/1998	210	210	<5	<5	240
20R04	11/17/1998	180	180	<5	<5	200
20R07	11/17/1998	210	210	<5	<5	240
21K05	11/17/1998	210	210	<5	<5	240
20R07	11/23/1998	-	-	-	-	-
28D12	2/24/1999	200	200	<5	<5	240
Mowry 1	2/24/1999	200	200	<5	<5	250
Mowry 7	2/24/1999	240	240	<5	<5	290
Mowry 7	3/9/1999	-	-	-	-	-

Sample I.D.	Date Sampled	Anions (mg/L)			
		Cl^-	SO_4^{-2}	NO_3^-	NO_2^-
Pit S at 5'*	4/16/1997	77	75	<0.10	<0.10
Pit S at 22'*	4/16/1997	76	75	<0.10	<0.10
Pit S at 52'*	4/16/1997	77	69	<0.10	<0.10
20J04	5/6/1997	84	-	<1.0	<1.0
20J06	5/6/1997	81	-	<1.0	<1.0
20J05	5/8/1997	86	-	1.0	<0.10
20J05†	5/8/1997	-	-	-	-
20R03	5/8/1997	82	-	1.8	<0.10
20R03†	5/8/1997	-	-	-	-
20R04	5/8/1997	72	-	4.2	<0.10

(continued)

Table B-2 (continued)

Sample I.D.	Date Sampled	Anions (mg/L)			
		Cl^-	SO_4^{-2}	NO_3^-	NO_2^-
20R04†	5/8/1997	-	-	-	-
20R05	5/8/1997	90	-	0.67	<0.10
20R05†	5/8/1997	-	-	-	-
20J04	6/18/1997	-	-	-	-
20J05	6/18/1997	-	-	-	-
20J06	6/18/1997	-	-	-	-
20R04	6/18/1997	-	-	-	-
20R05	6/18/1997	-	-	-	-
Bunting South at 3.3'	12/29/1997	95	90	7.4	<5
Bunting South at 3.3'†	12/29/1997	-	-	-	-
Bunting South at 16.4'	12/29/1997	95	89	7.4	<5
Bunting South at 16.4'†	12/29/1997	-	-	-	-
Bunting Middle at 3.3'	12/29/1997	95	90	7.9	<5
Bunting Middle at 3.3'†	12/29/1997	-	-	-	-
Bunting Middle at 32.8'	12/29/1997	95	89	7.5	<5
Bunting Middle at 32.8'†	12/29/1997	-	-	-	-
Kaiser AHF South at 3.3'	12/29/1997	91	86	7.3	<5
Kaiser AHF South at 3.3'†	12/29/1997	-	-	-	-
Kaiser AHF South at 36.1'	12/29/1997	91	85	7.3	<5
Kaiser AHF South at 36.1'†	12/29/1997	-	-	-	-
Kaiser AHF Middle at 3.3'	12/29/1997	92	89	7.4	<5
Kaiser AHF Middle at 3.3'†	12/29/1997	-	-	-	-
Kaiser AHF Middle at 62.3'	12/29/1997	94	84	8.2	<5
Kaiser AHF Middle at 62.3'†	12/29/1997	-	-	-	-
21L07	1/13/1998	98	81	<5	<5
21L08	1/13/1998	87	78	8.1	<5
21L05	2/4/1998	100	89	<5	<5
21L05‡	2/4/1998	-	-	-	-
21L06	2/4/1998	97	82	<5	<5
21L06‡	2/4/1998	-	-	-	-
21L07	2/4/1998	98	92	6.7	<5
21L07‡	2/4/1998	-	-	-	-
21K05	2/4/1998	90	77	<5	<5
21K05‡	2/4/1998	-	-	-	-
21L03	3/3/1998	98	100	2.1	<0.1
21L03†	3/3/1998	-	-	-	-
21J03	3/3/1998	87	84	0.94	<0.1
21J03†	3/3/1998	-	-	-	-
PT-7	4/7/1998	85	98	7.6	<1
PT-3	4/7/1998	90	98	8.3	<1
PT-1	6/9/1998	74	87	12	<1
PT-3	6/9/1998	75	83	5.4	<1
PT-6	6/9/1998	65	80	5.6	<1
PIT S Northeast at 10'	8/25/1998	66	88	<.05	<0.05

(continued)

Table B-2 (continued)

Sample I.D.	Date Sampled	Anions (mg/L)			
		Cl^-	SO_4^{-2}	NO_3^-	NO_2^-
PIT S Northeast at 40'	8/25/1998	54	71	<.05	<0.05
RD-3 Upstream-750' at 2'	8/25/1998	100	120	0.06	<0.05
RD-3 Upstream-Immediate	8/25/1998	110	120	<.05	<0.05
20J04	9/15/1998	93	96	0.1	<0.05
20R03	9/15/1998	87	100	<.05	<0.05
21H11	9/15/1998	54	2.3	<.05	<0.05
20J06	11/17/1998	75	86	0.31	<0.05
20R04	11/17/1998	61	79	0.20	<0.05
20R07	11/17/1998	74	86	0.27	<0.05
21K05	11/17/1998	82	90	0.24	<0.05
20R07	11/23/1998	-	-	-	-
28D12	2/24/1999	84	82	0.24	<0.05
Mowry 1	2/24/1999	80	85	0.76	<0.05
Mowry 7	2/24/1999	79	81	1.4	<0.05
Mowry 7	3/9/1999	-	-	-	-

Sample I.D.	Date Sampled	Cations (mg/L)				
		Ca^{+2}	Mg^{+2}	Na^+	K^+	B
Pit S at 5'*	4/16/1997	41	35	59	3.5	0.66
Pit S at 22'*	4/16/1997	40	31	57	3.3	0.60
Pit S at 52'*	4/16/1997	41	34	62	3.6	0.55
20J04	5/6/1997	49	36	79	3.3	0.67
20J06	5/6/1997	48	31	76	3.0	0.60
20J05	5/8/1997	46	24	65	1.6	0.72
20J05†	5/8/1997	-	-	-	-	-
20R03	5/8/1997	47	28	58	1.2	0.67
20R03†	5/8/1997	-	-	-	-	-
20R04	5/8/1997	44	23	62	<1.0	0.56
20R04†	5/8/1997	-	-	-	-	-
20R05	5/8/1997	50	23	69	<1.0	0.50
20R05†	5/8/1997	-	-	-	-	-
20J04	6/18/1997	-	-	-	-	-
20J05	6/18/1997	-	-	-	-	-
20J06	6/18/1997	-	-	-	-	-
20R04	6/18/1997	-	-	-	-	-
20R05	6/18/1997	-	-	-	-	-
Bunting South at 3.3'	12/29/1997	40	28	63	2	0.64
Bunting South at 3.3'†	12/29/1997	-	-	-	-	-
Bunting South at 16.4'	12/29/1997	37	26	57	1.8	0.62
Bunting South at 16.4'†	12/29/1997	-	-	-	-	-

(continued)

Table B-2 (continued)

Sample I.D.	Date Sampled	Cations (mg/L)				
		Ca^{+2}	Mg^{+2}	Na^{+}	K^{+}	B
Bunting Middle at 3.3'	12/29/1997	37	26	5.7	1.7	0.61
Bunting Middle at 3.3'†	12/29/1997	-	-	-	-	-
Bunting Middle at 32.8'	12/29/1997	38	26	58	1.8	0.62
Bunting Middle at 32.8'†	12/29/1997	-	-	-	-	-
Kaiser AHF South at 3.3'	12/29/1997	37	26	55	1.7	0.61
Kaiser AHF South at 3.3'†	12/29/1997	-	-	-	-	-
Kaiser AHF South at 36.1'	12/29/1997	38	26	55	1.7	0.6
Kaiser AHF South at 36.1'†	12/29/1997	-	-	-	-	-
Kaiser AHF Middle at 3.3'	12/29/1997	38	26	55	1.7	0.61
Kaiser AHF Middle at 3.3'†	12/29/1997	-	-	-	-	-
Kaiser AHF Middle at 62.3'	12/29/1997	37	25	53	1.6	0.58
Kaiser AHF Middle at 62.3'†	12/29/1997	-	-	-	-	-
21L07	1/13/1998	61	29	82	3.1	0.67
21L08	1/13/1998	50	37	79	2	0.57
21L05	2/4/1998	55	29	76	3.3	-
21L05‡	2/4/1998	-	-	-	-	-
21L06	2/4/1998	56	28	75	3.2	-
21L06‡	2/4/1998	-	-	-	-	-
21L07	2/4/1998	52	28	74	3.1	-
21L07‡	2/4/1998	-	-	-	-	-
21K05	2/4/1998	51	26	82	3.4	-
21K05‡	2/4/1998	-	-	-	-	-
21L03	3/3/1998	59	25	76	2.2	-
21L03†	3/3/1998	-	31	-	-	-
21J03	3/3/1998	62	25	76	2	-
21J03†	3/3/1998	-	-	-	-	-
PT-7	4/7/1998	57	30	81	2.5	-
PT-3	4/7/1998	65	30	84	2.1	-
PT-1	6/9/1998	78	34	88	2.3	0.94
PT-3	6/9/1998	60	27	76	2.8	0.73
PT-6	6/9/1998	53	27	73	3.2	0.69
PIT S Northeast at 10'	8/25/1998	27	22	44	1.9	0.48
PIT S Northeast at 40'	8/25/1998	37	22	45	2	0.47
RD-3 Upstream-750' at 2'	8/25/1998	38	28	67	2	0.67
RD-3 Upstream-Immediate	8/25/1998	38	28	65	2	0.67
20J04	9/15/1998	61	26	58	3.3	0.66
20R03	9/15/1998	55	26	62	3.4	0.66
21H11	9/15/1998	31	8.6	310	8.3	3.60
20J06	11/17/1998	52	26	70	3.8	0.70
20R04	11/17/1998	45	21	68	3.6	0.62
20R07	11/17/1998	51	27	75	3.9	0.67
21K05	11/17/1998	54	26	79	3.7	0.72
20R07	11/23/1998	-	-	-	-	-
28D12	2/24/1999	55	25	82	3.2	0.65

(continued)

Table B-2 (continued)

Sample I.D.	Date Sampled	Cations (mg/L)				
		Ca^{+2}	Mg^{+2}	Na^{+}	K^{+}	B
Mowry 1	2/24/1999	59	26	82	2.8	0.59
Mowry 7	2/24/1999	67	30	82	3.3	0.62
Mowry 7	3/9/1999	-	-	-	-	-

Sample I.D.	Date Sampled	Minor Metals and Nonionics (μg/L)				
		Cu	Fe	Mn	MBAS (mg/L)	Zn
Pit S at 5"*	4/16/1997	1.9	120	12	<0.025	9.1
Pit S at 22"*	4/16/1997	2.3	78	18	<0.025	13.0
Pit S at 52"*	4/16/1997	1.5	390	680	<0.025	13.0
20J04	5/6/1997	-	-	-	-	-
20J06	5/6/1997	-	-	-	-	-
20J05	5/8/1997	-	-	-	-	-
20J05†	5/8/1997	-	-	-	-	-
20R03	5/8/1997	-	-	-	-	-
20R03†	5/8/1997	-	-	-	-	-
20R04	5/8/1997	-	-	-	-	-
20R04†	5/8/1997	-	-	-	-	-
20R05	5/8/1997	-	-	-	-	-
20R05†	5/8/1997	-	-	-	-	-
20J04	6/18/1997	-	-	-	-	-
20J05	6/18/1997	-	-	-	-	-
20J06	6/18/1997	-	-	-	-	-
20R04	6/18/1997	-	-	-	-	-
20R05	6/18/1997	-	-	-	-	-
Bunting South at 3.3'	12/29/1997	2.4	87	4.3	<0.025	7.1
Bunting South at 3.3'†	12/29/1997	-	-	-	-	-
Bunting South at 16.4'	12/29/1997	2.3	78	4.7	<0.025	7.9
Bunting South at 16.4'†	12/29/1997	-	-	-	-	-
Bunting Middle at 3.3'	12/29/1997	1.9	73	4.2	0.029	9.0
Bunting Middle at 3.3'†	12/29/1997	-	-	-	-	-
Bunting Middle at 32.8'	12/29/1997	2	88	4.6	<0.025	7.0
Bunting Middle at 32.8'†	12/29/1997	-	-	-	-	-
Kaiser AHF South at 3.3'	12/29/1997	2.2	85	3.6	<0.025	5.4
Kaiser AHF South at 3.3'†	12/29/1997	-	-	-	-	-
Kaiser AHF South at 36.1'	12/29/1997	2	98	5.3	<0.025	9.3
Kaiser AHF South at 36.1'†	12/29/1997	-	-	-	-	-
Kaiser AHF Middle at 3.3'	12/29/1997	2.3	77	3.9	<0.025	4.8
Kaiser AHF Middle at 3.3'†	12/29/1997	-	-	-	-	-
Kaiser AHF Middle at 62.3'	12/29/1997	1.9	110	5.9	<0.025	6.0

(continued)

Table B-2 (continued)

Sample I.D.	Date Sampled	Minor Metals and Nonionics (μg/L)				
		Cu	Fe	Mn	MBAS (mg/L)	Zn
Kaiser AHF Middle at 62.3'†	12/29/1997	-	-	-	-	-
21L07	1/13/1998	1.4	45	660	<0.025	1.3
21L08	1/13/1998	1.2	25	68	<0.025	<0.4
21L05	2/4/1998	1.3	<5	520	<0.025	2.5
21L05‡	2/4/1998	-	-	-	-	-
21L06	2/4/1998	1.3	<5	570	<0.025	3.1
21L06‡	2/4/1998	-	-	-	-	-
21L07	2/4/1998	4.7	160	740	<0.025	6.4
21L07‡	2/4/1998	-	-	-	-	-
21K05	2/4/1998	3.1	<5	32	<0.025	2.4
21K05‡	2/4/1998	-	-	-	-	-
21L03	3/3/1998	1.7	<5	51	<0.025	3.7
21L03†	3/3/1998	-	-	1.4	-	-
21J03	3/3/1998	1.3	<5	1.3	<0.025	8.2
21J03†	3/3/1998	-	-	-	-	-
PT-7	4/7/1998	1.6	<10	5	0.031	1.9
PT-3	4/7/1998	1	<10	<0.4	<0.025	<0.8
PT-1	6/9/1998	16	510	77	<0.025	11.0
PT-3	6/9/1998	14	290	51	<0.025	10.0
PT-6	6/9/1998	6.5	69	12	<0.025	13.0
PIT S Northeast at 10'	8/25/1998	<5	20	6	<0.02	<10
PIT S Northeast at 40'	8/25/1998	<5	70	98	<0.02	<10
RD-3 Upstream-750' at 2'	8/25/1998	<5	40	5	<0.02	<10
RD-3 Upstream-Immediate	8/25/1998	<5	40	6	<0.02	<10
20J04	9/15/1998	<5	<10	<5	<0.02	<10
20R03	9/15/1998	<5	20	<5	<0.02	<10
21H11	9/15/1998	<5	2200	410	<0.02	<10
20J06	11/17/1998	<5	<10	28	<0.02	<10
20R04	11/17/1998	<5	<10	<5	<0.02	<10
20R07	11/17/1998	<5	30	<5	<0.02	<10
21K05	11/17/1998	<5	10	<5	<0.02	<10
20R07	11/23/1998	-	-	-	-	-
28D12	2/24/1999	<5	60	<5	<0.02	20.0
Mowry 1	2/24/1999	<5	50	<5	<0.02	30.0
Mowry 7	2/24/1999	<5	80	<5	<0.02	20.0
Mowry 7	3/9/1999	-	-	-	-	-

Sample I.D.	Date Sampled	Solids	Org. Carbon	Biological	Field Parameters		
		TDS (mg/L)	TOC (mg/L)	MS Coliphage (pfu/100 ml)	Turbidity (NTU)	Measured By Lab Conductivity (μmhos/cm)	pH
Pit S at 5'*	4/16/1997	400	4.7	-	-	710	8.5

(continued)

Table B-2 (continued)

Sample I.D.	Date Sampled	Solids: TDS (mg/L)	Org. Carbon: TOC (mg/L)	Biological: MS Coliphage (pfu/100 ml)	Field Parameters: Turbidity (NTU)	Field Parameters (Measured By Lab): Conductivity (µmhos/cm)	Field Parameters: pH
Pit S at 22'*	4/16/1997	370	4.4	-	-	700	8.1
Pit S at 52'*	4/16/1997	390	6.4	-	-	730	7.7
20J04	5/6/1997	470	2.1	-	-	-	-
20J06	5/6/1997	470	1.2	-	-	-	-
20J05	5/8/1997	440	7.8	-	-	-	-
20J05†	5/8/1997	-	<5.0	-	-	-	-
20R03	5/8/1997	460	9.9	-	-	-	-
20R03†	5/8/1997	-	<5.0	-	-	-	-
20R04	5/8/1997	430	9.3	-	-	-	-
20R04†	5/8/1997	-	<5.0	-	-	-	-
20R05	5/8/1997	450	8.2	-	-	-	-
20R05†	5/8/1997	-	<5.0	-	-	-	-
20J04	6/18/1997	-	1.3	-	-	-	-
20J05	6/18/1997	-	1.0	-	-	-	-
20J06	6/18/1997	-	<1.0	-	-	-	-
20R04	6/18/1997	-	1.5	-	-	-	-
20R05	6/18/1997	-	1.4	-	-	-	-
Bunting South at 3.3'	12/29/1997	480	5.6	3	-	820	8.4
Bunting South at 3.3'†	12/29/1997	-	20	-	-	-	-
Bunting South at 16.4'	12/29/1997	420	19	3	-	760	8.4
Bunting South at 16.4'†	12/29/1997	-	9	-	-	-	-
Bunting Middle at 3.3'	12/29/1997	490	19	<1	-	820	8.5
Bunting Middle at 3.3'†	12/29/1997	-	7.6	-	-	-	-
Bunting Middle at 32.8'	12/29/1997	540	18	1	-	850	8.5
Bunting Middle at 32.8'†	12/29/1997	-	13	-	-	-	-
Kaiser AHF South at 3.3'	12/29/1997	560	53	<1	-	810	8.4
Kaiser AHF South at 3.3'†	12/29/1997	-	<5	-	-	-	-
Kaiser AHF South at 36.1'	12/29/1997	490	30	<1	-	860	8.4
Kaiser AHF South at 36.1'†	12/29/1997	-	6.3	-	-	-	-
Kaiser AHF Middle at 3.3'	12/29/1997	500	28	<1	-	810	8.4
Kaiser AHF Middle at 3.3'†	12/29/1997	-	14	-	-	-	-
Kaiser AHF Middle at 62.3'	12/29/1997	470	25	7	-	820	8.4
Kaiser AHF Middle at 62.3'†	12/29/1997	-	<5	-	-	-	-
21L07	1/13/1998	530	11	<1	-	880	7.7
21L08	1/13/1998	460	11	<1	-	820	7.7
21L05	2/4/1998	490	7.3	<1	-	800	7.5
21L05‡	2/4/1998	-	1.4	-	-	-	-
21L06	2/4/1998	490	32	<1	-	800	7.6
21L06‡	2/4/1998	-	1.5	-	-	-	-
21L07	2/4/1998	500	14	<1	-	810	7.4
21L07‡	2/4/1998	-	1.7	-	-	-	-
21K05	2/4/1998	510	29	<1	-	770	8
21K05‡	2/4/1998	-	1.3	-	-	-	-

(continued)

Table B-2 (continued)

Sample I.D.	Date Sampled	Solids	Org. Carbon	Biological	Field Parameters		
		TDS (mg/L)	TOC (mg/L)	MS Coliphage (pfu/100 ml)	Turbidity (NTU)	Measured By Lab Conductivity (μmhos/cm)	pH
21L03	3/3/1998	502	23	<1	-	850	7.4
21L03†	3/3/1998	-	-	-	-	-	-
21J03	3/3/1998	485	21	<1	-	830	7.4
21J03†	3/3/1998	-	1.5	-	-	-	-
PT-7	4/7/1998	510	<5	<1	-	880	7.8
PT-3	4/7/1998	550	<5	<1	-	930	8
PT-1	6/9/1998	610	1.5	<1	-	870	7.4
PT-3	6/9/1998	490	6	<1	-	710	7.6
PT-6	6/9/1998	510	3.6	<1	-	710	7.6
PIT S Northeast at 10'	8/25/1998	370	<1	<1	-	740	8.7
PIT S Northeast at 40'	8/25/1998	400	<1	<1	-	710	8.1
RD-3 Upstream-750' at 2'	8/25/1998	550	<1	<1	-	990	8.5
RD-3 Upstream-Immediate	8/25/1998	570	<1	<1	-	1000	8.4
20J04	9/15/1998	530	<1	<1	-	930	7.6
20R03	9/15/1998	520	<1	<1	-	880	7.9
21H11	9/15/1998	910	<1	<1	-	1500	7.2
20J06	11/17/1998	480	<1	<1	-	840	7.6
20R04	11/17/1998	430	<1	<1	-	740	7.5
20R07	11/17/1998	480	<1	-	-	830	7.4
21K05	11/17/1998	510	<1	<1	-	870	7.5
20R07	11/23/1998	-	-	<1	-	-	-
28D12	2/24/1999	500	<1	<1	-	840	7.4
Mowry 1	2/24/1999	520	<1	<1	-	870	7.3
Mowry 7	2/24/1999	530	<1	15	-	920	7.4
Mowry 7	3/9/1999	-	-	<1	-	-	-

* Although not included in this report, lab and field data sheets label these Pit S samples at 8', 25', and 55', respectively
Difference due to initial measuring point being 3 ft. higher than water surface.

† Sample was re-analyzed by the contract laboratory.

‡ Analyzed by ACWD's in-house water quality laboratory.

Table B-3

MTBE and TOC (analyzed at LLNL)

Sample ID	Sub-basin	Date	MTBE (ppt)	TOC (mg/L)
21J03	AHF	3/3/97	20	--
21J03	AHF	12/16/97	31	--
21L08	AHF	12/16/97	72	--
Bunting Middle at 16'	AHF	12/16/97	135	--
Bunting West at 4.9'	AHF	12/29/97	193	--
Bunting West at 20'	AHF	12/29/97	142	--
Kaiser AHF East at 3'	AHF	12/29/97	133	--
Kaiser AHF East at 24'	AHF	12/29/97	116	--
Kaiser AHF South at 2'	AHF	12/29/97	142	--
Kaiser AHF South at 22.5'	AHF	12/29/97	127	--
Rain Water in Livermore, Calif.	Outside ACWD	1/10/98	67	--
21K05	AHF	2/4/98	77	--
21L07	AHF	2/4/98	76	--
21L05	AHF	2/4/98	15	--
21L03 - Upper Zone	AHF	3/3/98	ND	--
PT-3	AHF	3/3/98	ND	--
PT-7	AHF	3/3/98	20	--
PT-1	AHF	6/9/98	ND	--
PT-3	AHF	6/9/98	ND	--
PT-6	AHF	6/9/98	ND	--
Pit S Northeast at 5'	BHF	8/25/98	84	3.4
Pit S Northeast at 23'	BHF	8/25/98	95	2.8
Pit S South at 5'	BHF	8/25/98	108	--
Pit S South at 23'	BHF	8/25/98	97	3.1
20R03	BHF	9/15/98	97	1
21H11	AHF	9/15/98	52	0.4
20J04	BHF	9/15/98	90	0.8
20J06	BHF	11/17/98	30	0.76
20R04	BHF	11/17/98	29	0.76
20R07	BHF	11/17/98	ND	0.56
21K05	AHF	11/17/98	98	0.74
RD3 Upstream-750' at 5'	AHF	11/20/98	213	1.6

Table B-4

Concentrations of trace metals (principal isotope) in Kaiser Pond-AHF and nearby AHF wells

Element	Kaiser at 2' 12/29/97 South Station	Kaiser at 22.5' 12/29/97 South Station	Kaiser at 3' 12/29/97 East Station	21L08 12/16/97	21L07 12/2/97	21L07 2/4/98	21L05 12/2/97	21L05 2/4/98	21L06 12/2/97	21L06 2/4/98	21L03 3/3/98 Upper Screen
Al27	65.5	122.2	143.2	--	386.3	22.0	53.8	--	424.87	--	--
Rb85	0.6	0.8	0.6	0.3	0.9	0.3	0.6	0.3	1.22	0.32	0.4
Sr88	758.7	772.8	757.7	717.2	700.8	769.9	820.6	824.0	816.42	795.56	981.0
Mo98	3.5	3.7	3.3	2.8	4.3	3.7	2.6	2.2	3.52	3.31	0.5
Cd114	0.07	0.09	0.07	0.02	0.07	0.05	0.02	0.03	--	0.08	0.08
Sb121	0.13	0.16	0.13	0.07	0.06	0.07	0.01	0.05	0.15	0.11	ND
Ba138	116.9	113.3	114.3	93.0	107.2	122.3	113.8	124.4	137.11	135.53	140.0
W184	ND	ND	ND	ND	2.7	ND	0.1	ND	0.02	--	ND
Pb208	0.53	0.53	0.28	0.98	0.69	0.53	0.41	0.99	1.4	0.52	2.36
U238	2.1	2.2	2.1	2.3	1.1	1.2	1.0	0.8	0.93	0.87	0.5
V51	2.5	3.4	3.2	2.5	2.4	2.7	1.2	1.3	2.96	2.07	0.9
Cr52	1.3	1.4	1.5	1.7	2.6	1.2	1.0	1.5	3.74	1.07	1.4
Mn55	3.9	4.5	4.7	1.2	642.5	682.4	503.7	496.8	549.34	540.9	1.6
Fe56	93.2	101.2	108.2	9.3	269.8	3.8	11.6	5.9	768.12	7.78	3.2
Co59	ND	0.3	0.4	0.7	0.6	1.0	0.1	0.7	0.74	0.45	ND
Ni60	1.4	1.7	1.5	--	1.4	1.7	1.3	--	2.69	0.36	--
Cu63	2.2	3.4	3.2	6.3	3.7	4.2	1.4	2.9	2.75	2.55	4.3
Zn66	7.0	12.0	9.2	23.4	11.9	9.3	12.6	7.5	1.272	5.61	16.6

Note: all concentrations in μg/L.

Table B-5

Concentrations of trace metals (principal isotope) in Bunting Pond and nearby well 21K05

Element	Bunting at 3' 12/29/97 Southeast	Bunting at 15' 12/29/97 Southeast	21K05 2/4/98
Al27	161.96	103.43	31.60
Rb85	0.67	0.50	0.45
Sr88	779.39	737.65	744.92
Mo98	3.93	3.66	3.68
Cd114	0.03	0.03	ND
Sb121	0.24	0.24	0.12
Ba138	128.35	121.63	107.64
W184	ND	ND	ND
Pb208	0.69	0.52	0.70
U238	2.45	2.29	1.39
V51	3.99	3.8	1.19
Cr52	1.36	1.42	1.02
Mn55	4.99	4.34	30.25
Fe56	90.84	80.22	55.12
Co59	0.84	ND	0.95
Ni60	1.32	1.14	15.30
Cu63	3.16	7.75	2.45
Zn66	9.00	15.17	16.18

Note: all concentrations are in μg/L.

Table B-6

Concentrations of trace metals in Alameda Creek (upstream of Rubber Dam No. 3), Pit S, and nearby wells

Element	RD3 upstream- approx 750' at 2' 8/25/98 (AHF)	RD3 upstream- immediate at 2' 8/25/98 (AHF)	RD3 upstream- immediate at 6' 8/25/98 (AHF)	Pit S south at 40' 8/25/98 (BHF)	Pit S northeast at 23' 8/25/98 (BHF)	Pit S northeast at 5' 8/25/98 (BHF)	21H11 9/15/98 (AHF)	20J04 9/15/98 (BHF)	20R03 9/15/98 (BHF)
Al27	41.32	33.71	34.11	7.91	0.85	0.99	ND	ND	7.62
Mo98	4.54	4.66	5.08	2.69	3.15	3.26	1.10	2.95	3.70
Cd114	ND	0.02	ND	0.08	ND	ND	ND	0.01	ND
Pb208	0.17	0.06	0.06	0.04	0.05	0.03	<.03	<.03	<.03
U238	8.75	8.45	8.54	5.01	5.65	6.02	0.03	4.28	4.47
V51	3.61	3.14	3.01	1.94	2.65	3.50	0.96	2.19	2.72
Cr52	3.22	2.02	2.09	1.57	1.24	1.71	3.02	1.55	3.76
Mn55	5.82	3.18	3.05	102.60	0.48	3.56	612.00	0.35	2.90
Fe56	873.90	244.70	239.10	341.50	116.50	176.00	1527.00	142.70	186.50
Co59	0.43	1.25	1.28	0.26	0.24	1.10	0.27	0.19	0.15
Ni60	10.39	4.89	5.59	5.39	2.57	3.69	0.90	1.50	1.54
Cu63	5.45	5.98	5.80	3.24	3.63	3.94	0.23	5.50	5.70
Zn68	11.49	8.82	8.71	5.23	1.66	1.31	87.60	9.40	8.48
As75	2.92	3.03	2.93	2.61	2.25	2.96	3.01	0.96	1.21
Ag107	0.01	0.01	0.02	0.01	0.01	0.01	0.02	0.01	0.01

Note: all concentrations are in μg/L.

REFERENCES

ACWD (Alameda County Water District). 1997. *Groundwater Monitoring Report, Fall 1997.* Fremont, Calif.: ACWD

Aeschbach-Hertig W., F. Peeters, U. Beyerle, and R. Kipfer. 1999. Interpretation of Dissolved Atmospheric Noble Gases in Natural Waters, *Water Resources Research*, 35(9): 2779-2792.

Ahel, M. 1991. Infiltration of Organic Pollutants into Groundwater: Field Studies in the Alluvial Aquifer of the Sava River. *Bulletin of Environmental Contamination and Toxicology*, 47: 586-593.

Bales, R.C., S. Li., K.M. Maguire, M.T. Yahya, and C.P. Gerba. 1993. MS-2 and Poliovirus Transport in Porous Media: Hydrophobic Effects and Chemical Perturbations. *Water Resources Research,* 29(4): 957-963.

Bales, R.C., S. Li, K.M. Maguire, M.T. Yahya, C. P. Gerba, and R.W. Harvey. 1995. Virus and Bacteria Transport in a Sandy Aquifer, Cape Cod, Mass. *Ground Water,* 33(4): 653-661.

Banks, H.O. 1979. Field Testing of the Alameda County Water District Kaiser Pit Recharge Facility April 17 – May 5, 1979. Report to Alameda County Water District, December 20, 1979. H.O. Banks (a Camp, Dresser, and McKee firm), Belmont, Calif.

Bouwer, H. 1994. Groundwater Recharge Using Waters of Impaired Quality. National Academy of Sciences. http://www.nap.edu/openbook/0309051428/html/3.html.

Brown, S.L. 1997. Atmospheric and Potable Water Exposures to Methyl-tert-Butyl Ether (MTBE). *Regulatory Toxicology and Pharmacology*, 25(3): 256-276.

Chapra, S.C. and K.H. Reckhow. 1983. Engineering Approaches for Lake Management, Volume 2: Mechanistic Modeling. Boston: Butterworth Publishers.

Craig, H. 1961. Isotopic Variations in Meteoric Waters. *Science*, 133: 1702-1703.

Criss, R. E. 1999. *Principles of Stable Isotope Distribution.* New York: Oxford University Press.

Davisson, M.L., G. B. Hudson, J.E. Moran, S. Niemeyer, and R. Herndon. 1998. Isotope Tracer Approaches for Characterizing Artificial Recharge and Demonstrating Regulatory Compliance. In *Proc. of the Annual UC Water Reuse Research Conference.* Monterey, Calif.

DWR (State of California Department of Water Resources). 1960. Bulletin No. 81. *Intrusion of Salt Water into Groundwater Basins of Southern Alameda County.* Sacramento, Calif.: DWR.

DWR (State of California Department of Water Resources). 1963. Bulletin No. 13: *Alameda County Investigation.* Sacramento, Calif.: DWR.

DWR (State of California Department of Water Resources). 1967. Bulletin No. 118-1. *Evaluation of Groundwater Resources, South Bay, Appendix-A: Geology.* Sacramento, Calif.: DWR.

DWR (State of California Department of Water Resources). 1968. Bulletin No. 118-1: *Evaluation of Ground Water Resources, South Bay, Volume I: Fremont Study Area.* Sacramento, Calif.: DWR.

Drever, James. 1988. *Geochemistry of Natural Waters.* Englewood Cliffs, N.J.: Prentice Hall.

Drewes, J.E., and P. Fox. 1999a. Behavior and Characterization of Residual Organic Compounds in Wastewater Used for Indirect Potable Reuse. *Water Science and Technology*, 40(4-5): 391-398.

Drewes, J.E., and P. Fox. 1999b. Fate of Natural Organic Matter (NOM) During Groundwater Recharge Using Reclaimed Water. *Water Science and Technology*, 40(9): 241-248.

Drewes, J.E., R. Rauch, M. Rincon, M.H. Nellor, and P. Fox. 1999. A Watershed-Guided Approach for Water Quality Assurance in Indirect Potable Reuse Systems: Experiences From Field Studies in Arizona and California. In *Proc. of the Annual Water Reuse Research Conference*, Monterey, Calif.

Ekwurzel, B., J.E. Moran, C.J. Koester, M.L. Davisson, and G.F. Eaton. 2001. Non-point Source MTBE Movement Through the Environment: Ultra Low-Level (ppt) Measurements in California. (17-26). In *ACS Symposium Series 799, Oxygenates in Gasoline Environmental Aspects.* Edited by A.F. Diaz and D.L. Drogos. Oxford University Press.

Erickson, B.E. 2001. Potential Aerobic MTBE Biodegradation. *Environmental Science & Technology, A Pages,* 35(5): 96A-97A.

Gat, J.R. 1980. The Isotopes of Hydrogen and Oxygen in Precipitation (21-47). In *Handbook of Environmental Isotope Geochemistry.* Edited by P. Fritz and J.C. Fontes. Amsterdam: Elsevier Science.

Hackley, K.C., C.L. Liu, and D.D. Coleman. 1996. Environmental Isotope Characteristics of Landfill Leachates and Gases. *Ground Water*, 34(5): 827-836.

Hamilton-Taylor, J. and W. Davison. 1995. Redox-Cycling of Trace Elements in Lakes. In *Physics and Chemistry of Lakes*, 2nd ed. Edited by A. Lerman, D. M. Imboden, and J. R. Gat). New York: Springer-Verlag.

Hessen, D.O., E.T. Gjessing, J. Knulst, and E. Fjeld. 1997. TOC Fluctuations in a Humic Lake as Related to Catchment Acidification, Season, and Climate. *Biogeochemistry*, 36: 139-151.

Hoefs, J. 1980. *Stable Isotope Geochemistry*, 2nd edition. New York: Springer-Verlag.

Holloway, J.M., R.A. Dahlgren, B. Hansen, and W.H. Casey. 1998. Contribution of Bedrock Nitrogen to High Nitrate Concentrations in Stream Water. *Nature*, 395(6704): 785-788.

Hudson, G., B. Davisson, M.L. Velsko, C.A. Niemeyer, S. Esser, K. Bradley, and J. Beiriger. 1995. Preliminary Report on Isotope Abundance Measurements in Groundwater Samples from the Talbert Injection Barrier Area, Orange County Water District. Livermore, Calif.: Lawrence Livermore National Laboratory.

International Consultants. 1998. Groundwater Rule Vulnerability Assessment Study, Revision 2 (Final Report). International Consultants, Inc., Dayton, Ohio. USEPA Contract No. 68-C6-0039. Washington, D.C.: USEPA.

Jensen, D.L, J.K. Boddum, S. Redemann, and T.H. Christensen. 1998. Speciation of Dissolved Iron(II) and Manganese(II) in a Groundwater Pollution Plume. *Environmental Science and Technology*, 32(18): 2657-2664.

Jin, Y., M.V. Yates, S.S. Thompson, and W.A. Jury. 1997. Sorption of Viruses during Flow through Saturated Sand Columns. *Environmental Science and Technology*, 31(2): 548-555.

Johnson, J.S., L.A. Baker, and P. Fox. 1999. Geochemical Transformations during Artificial Groundwater Recharge: Soil-Water Interactions of Inorganic Constituents. *Water Research*, 33(1): 196-206.

Kolterman, C. 1993. Geologic Modeling of Spatial Variability in Sedimentary Environments for Groundwater Flow Simulation. Ph.D. diss., Stanford University, Palo Alto, Calif.

Lienert, C., S.A. Short, and H.R. von Genten. 1994. Uranium Infiltration from a River to Shallow Groundwater. *Gechochimica et Cosmochimica Acta*, 58(24): 5455-5463.

Luhdorff and Scalmanini Consulting Engineers. 2002. Memorandum, Results of Geologic Investigation, East Bay Plain-Niles Cone Aquifer Test. May 21, 2002. Luhdorff and Scalmanini Consulting Engineers, Woodland, Calif.

McKay, L. D., J. A. Cherry, R. C. Bales, M. T. Yahya, and C. Gerba. 1993. A Field Example of Bacteriophage as Tracers of Fractured Flow. *Environmental Science and Technology*, 27: 1075-1079.

Montgomery Watson. 1996. Technical Memorandum to Alameda County Water District: Groundwater Quality Model Development. Montgomery Watson, Inc., Sacramento, Calif.

Mook, W.G. 1980. Carbon-14 in Hydrogeological Studies. In *Handbook of Environmental Isotope Geochemistry* 1A: 49-74. Edited by P. Fritz and J. C. Fontes. Amsterdam: Elsevier Science.

Pankow, J.F., N.R. Thomson, R.L. Johnson, A.L. Baehr, and J.S. Zogorski. 1997. The Urban Atmosphere as a Non-Point Source for the Transport of MtBE and Other Volatile Organic Compounds (VOCs) to Shallow Groundwater. *Environmental Science and Technology*, 31: 2821-2828.

Pieper, A. P., J. N. Ryan, R. W. Harvey, G. L. Amy, T. H. Illangasekare, and D. W. Metge. 1997. Transport and Recovery of Bacteriophage PRD1 in an Unconfined Sand Aquifer: Effect of Sewage-Derived Organic Matter. *Environmental Science and Technology*, 31: 1163-1170.

Rehmann, L.L.C., C.W. Welty, and R.W. Harvey. 1999. Stochastic Analysis of Virus Transport in Aquifers. *Water Resources Research*, 35(7): 1987-2006.

Roberts, P. V., J. Schreiner, and G. D. Hopkins. 1982. Field Study of Organic Water Quality Changes during Groundwater Recharge in Palo Alto Baylands. *Water Research*, 16: 1025-1035.

Salbu B. and E. Steinnes. 1995. *Trace Elements in Natural Waters*. Boca Raton, Fla.: CRC Press.

Schirmer, M., B.J. Butler, J.F. Barker, C.D. Church, and K. Schirmer. 1999. Evaluation of Biodegradation and Dispersion as Natural Attenuation Processes of MTBE and Benzene at the Borden Field Site. *Physics and Chemistry of the Earth Part B- Hydrology Oceans and Atmosphere*, 24(6): 557-560.

Schlosser, P., M. Stute, H. Dorr, C. Sonntag, and K.O. Munnich. 1988. Tritium/^{3}He Dating of Shallow Groundwater. *Earth Planet. Sci. Lett.*, 89: 353-362.

Schlosser, P., M. Stute, H.C. Sonntag, and K.O. Munnich. 1989. Tritiogenic ^{3}He in Shallow Groundwater. *Earth Planet. Sci. Lett.*, 94: 245-256.

Scholl, M.A., and R.W. Harvey. 1992. Laboratory Investigations on the Role of Sediment Surface and Groundwater Chemistry in Transport of Bacteria Through a Contaminated Sandy Aquifer. *Environmental Science Technology*, 26:1410-1417.

Schwarzenbach, R. P., W. Giger, E. Hoehn, and J. K. Schneider. 1983. Behavior of Organic Compounds during Infiltration of River Water to Groundwater. Field Studies. *Environmental Science and Technology*, 17: 472-479.

Squillace, P.J., J.F. Pankow, N.E. Korte, and J.S. Zogorski. 1996. *Environmental Behavior and Fate of Methyl tert-Butyl Ether (MtBE).* U.S. Geological Survey Fact Sheet, FS-203-96.

Stute, M., and P. Schlosser. 1993. Principles and Applications of the Noble Gas Paleothermometer. In *Climate Change in Continental Isotopic Records, Geophysical Monograph Series*, 78: 89-100. Edited by P.K. Swart, K.C. Lohmann, J. McKenzie, and S. Savin. Washington, D.C.: AGU.

Thompson, A.F.B, S.F. Carle, N.D. Rosenberg, and R.M. Maxwell. 1999. Analysis of Groundwater Migration from Artificial Recharge in a Large Urban Aquifer: A Simulation Perspective. *Water Resources Research*, 35(10): 2981-2998.

Toran, L. and A. V. Palumbo. 1992. Colloid Transport through Fractured and Unfractured Laboratory Sand Columns. *Journal of Contaminant Hydrology*, 9: 289-303.

USEPA (U.S. Environmental Protection Agency, Office of Research and Development). 1998. *Oxygenates in Water: Critical Information and Research Needs.* EPA/600/R-98/048.

Von Gunten, H. R., G. Karametaxas, U. Krähenbühl, M. Kuslys M., R. Giovanoli, E. Hoehn, and R. Keil. 1991. Seasonal Biogeochemical Cycles in Riverborne Groundwater. *Geochimica et Cosmochimica Acta* 55(12): 3597-3609.

Wigley, T.M.L. 1975. Carbon-14 Dating of Groundwater from Closed and Open Systems. *Water Resources Research*, 11: 324-328.

Williams, A.E., L.J. Lund, J.A. Johnson, and Z.J. Kabala. 1998. Natural and Anthropogenic Nitrate Contamination of Groundwater in a Rural Community, California. *Environmental Science & Technology*, 32(1): 32-29.

Woessner, W.W., P.N. Patrick, N. Ball, D.C. DeBorde, and T.L. Troy. 2001. Viral Transport in a Sand and Gravel Aquifer under Field Pumping Conditions. *Groundwater*, 39 (6): 886-894.

Yates, M.V., S.R. Yates, J. Wagner and C.P. Gerba. 1987. Modeling Virus Survival and Transport in the Subsurface. *Journal of Contaminant Hydrology*, 1 (1987): 329-345.

ABBREVIATIONS LIST

ACWD	Alameda County Water District
AHF	Above Hayward Fault (aquifer or sub-basin)
ARP	Aquifer Reclamation Program
AWWA	American Waterworks Association
AwwaRF	Awwa Research Foundation
BHF	Below Hayward Fault (sub-basin)
b.g.s.	below ground surface
°C	degrees centigrade
cc	cubic centimeter
CF Aquifer	Centerville-Fremont Aquifer
cfs	cubic feet per second
cm^3	cubic centimeter
D	deuterium (^2H)
DIC	dissolved inorganic carbon
D.O.	dissolved oxygen
DTW	depth to water (inside a well)
DWR	State of California Department of Water Resources
E_h	oxidation-reduction potential
ft	feet
GC-MS	gas chromatography - mass spectroscopy
IGSM	Integrated Groundwater Surface Water Model
in	inches
g	gram
gpm	gallons per minute
IGSM	Integrated Groundwater Surface Water Model (Montgomery Watson 1996)
L	liter
LLNL	Lawrence Livermore National Laboratory
MBAS	methylene-blue active substances
mg	milligram

ml	milliliter
MPE	measuring point elevation (of a well at ground surface)
MS	mass spectroscopy
	Male-Specific (Coliphage)
MSL	Mean Sea Level
MTBE	Methyl Tertiary Butyl Ether
MWL	Meteoric Water Line
mV	millivolts
NTU	nephelometric turbidity units
ORP	oxidation-reduction potential
pCi/L	pico-curies per liter
pmC	percent modern carbon
ppb	part per billion
ppm	part per million
ppt	part per trillion
ORP	oxidation-reduction potential
pCi/L	pico-curies per liter
pH	negative logarithm of the effective hydrogen ion concentration
pfu	plaque-forming units
ppt	part per trillion
PT	Peralta-Tyson (wells or wellfield)
PVC	polyvinyl chloride
RD	Rubber Dam (as in Rubber Dam No. 1)
Redox	oxidation-reduction potential
SF Bay	San Francisco Bay
STP	standard temperature and pressure
SWL	surface water elevation
TDS	total dissolved solids
temp	temperature
TOC	total organic carbon
µg	microgram
µmhos/cm	micromho per centimeter

vol.	volume
yr	year

IWA Publishing's authorised EU representative for General Product Safety Regulations is Diane D'Arras, 15 rue Duret, 75116 Paris, France, e-mail: safety@iwap.co.uk.

Printed and bound by CPI Group (UK) Ltd, Croydon, CR0 4YY
17/09/2026
02218808-0003